Maths — No Problem!

English National Curriculum

Consultant Editors
Dr Wong Khoon Yoong
Dr Julie Alderton

Series Editors
Adam Gifford
Andy Psarianos

Authors
Adam Gifford
Judy Hornigold

MATHS NO PROBLEM!

Published by Maths — No Problem!

Printed in the United Kingdom
First Printing, 2022

ISBN 978-1-913458-66-9

Maths — No Problem!
hello@mathsnoproblem.com
www.mathsnoproblem.com

Acknowledgements
The publisher would like to thank Dr Anne Hermanson and Alex Laurie for their contributions.

Preface

Welcome to Maths — No Problem!

Maths — No Problem! is a comprehensive series of textbooks and workbooks supported by detailed online teacher guidance to deliver maths mastery in primary schools. The materials use a spiral design with carefully built-up mathematical concepts and processes adapted from the maths mastery approaches used in Singapore. An integral part of the learning process is the Concrete, Pictorial, Abstract (CPA) approach, which forms the backbone of the materials developed for this series.

Maths — No Problem! incorporates the use of concrete aids and manipulatives, problem solving, group work, individual practice and journaling.

Meet our Maths — No Problem! friends who will join you on your journey through this textbook.

The Maths — No Problem! approach is exemplified throughout the chapters by the following teaching elements:

Chapter Opener

The chapter opener depicts familiar events or contexts that serve as an introduction for pupils to the concept covered in the chapter. It can also be used as an additional prompt for journal writing or an additional exercise for pupils.

Explore

Explore is an introductory activity for pupils, which includes questions related to various lesson objectives.

Master

Each lesson contains a Master section that develops lesson objectives by introducing new concepts through a CPA approach using manipulatives and engaging pictures. Guided examples are provided to reinforce the learning objective.

Activity Time

Activity Time provides pupils with opportunities to work as individuals or in small groups to explore mathematical concepts through games and investigations.

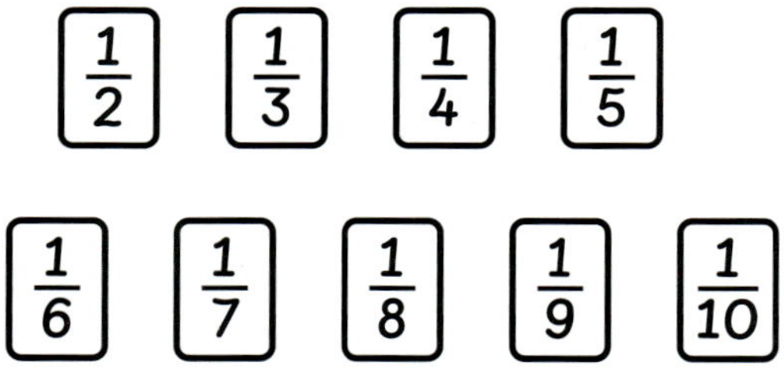

Guided Practice

Guided Practice contains questions for further consolidation and the immediate evaluation of pupils' learning.

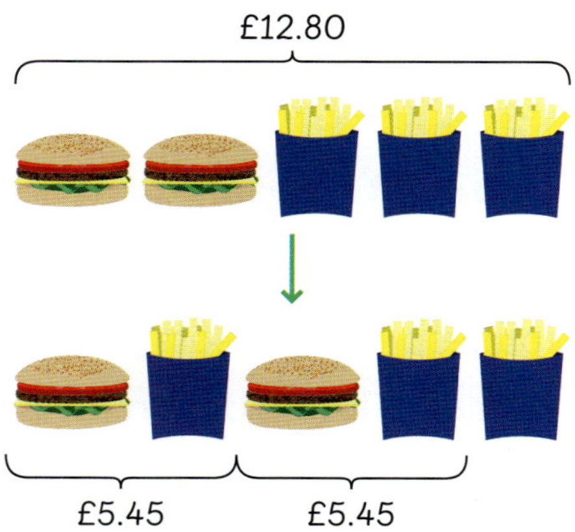

Mind Challenge

Mind Challenges are non-routine questions for pupils to apply relevant problem-solving skills and develop higher-order thinking skills.

My Maths Journal

My Maths Journal provides pupils with opportunities to show their understanding of the mathematical concepts learnt and to record their learning.

255 roses	a dozen roses
packs	boxes

Review

Review allows pupils to reflect on their own learning after each chapter, with support from the teacher.

Review

I can...

- ☐ convert measurements of length.
- ☐ convert measurements of mass.
- ☐ convert measurements of volume.
- ☐ convert measurements of time.

Contents

Chapter 1 — Numbers to 10 Million

Page

Chapter 2 — Four Operations on Whole Numbers

Page

Paint 2.79 l
Paint 2.79 l
Paint 2.79 l
Paint 2.79 l
Paint 2.79 l
Paint 2.79 l

Page

Page

Page

Chapter 1
Numbers to 10 Million

Reading and Writing Numbers to 10 Million

Lesson 1

Explore

The population of Lithuania is 2 713 406.
What are the different ways we can show this number using place value?

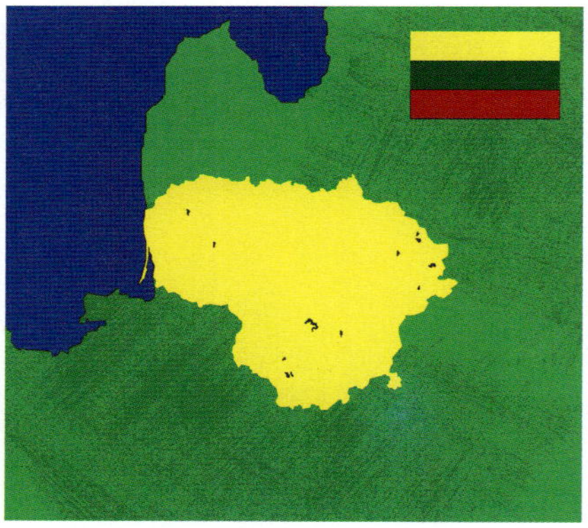

Master

Method 1

Use place-value counters to show 2 713 406.

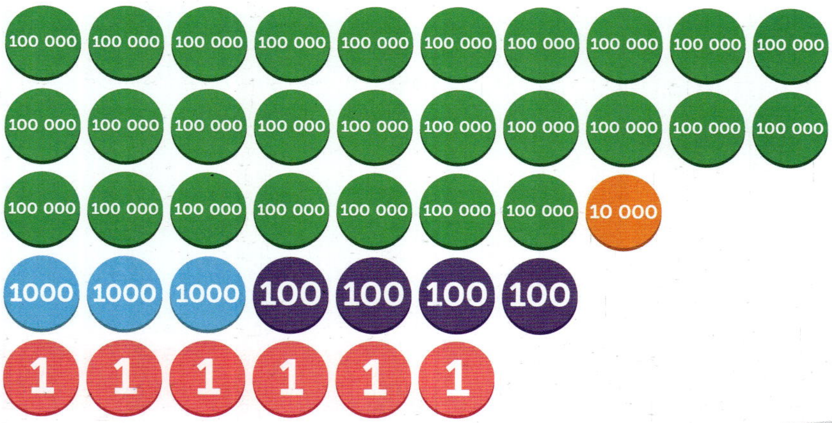

Method 2

Use a place-value chart to show 2 713 406.

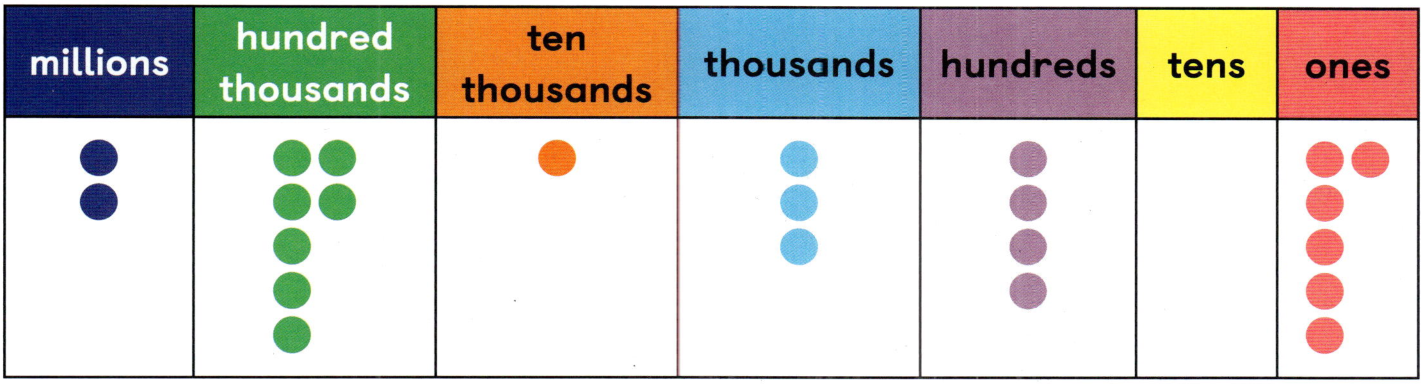

millions	hundred thousands	ten thousands	thousands	hundreds	tens	ones
2	7	1	3	4	0	6

Method 3

2 713 406 =

______ + ______ + ______ + ______ + ______ + ______ + ______

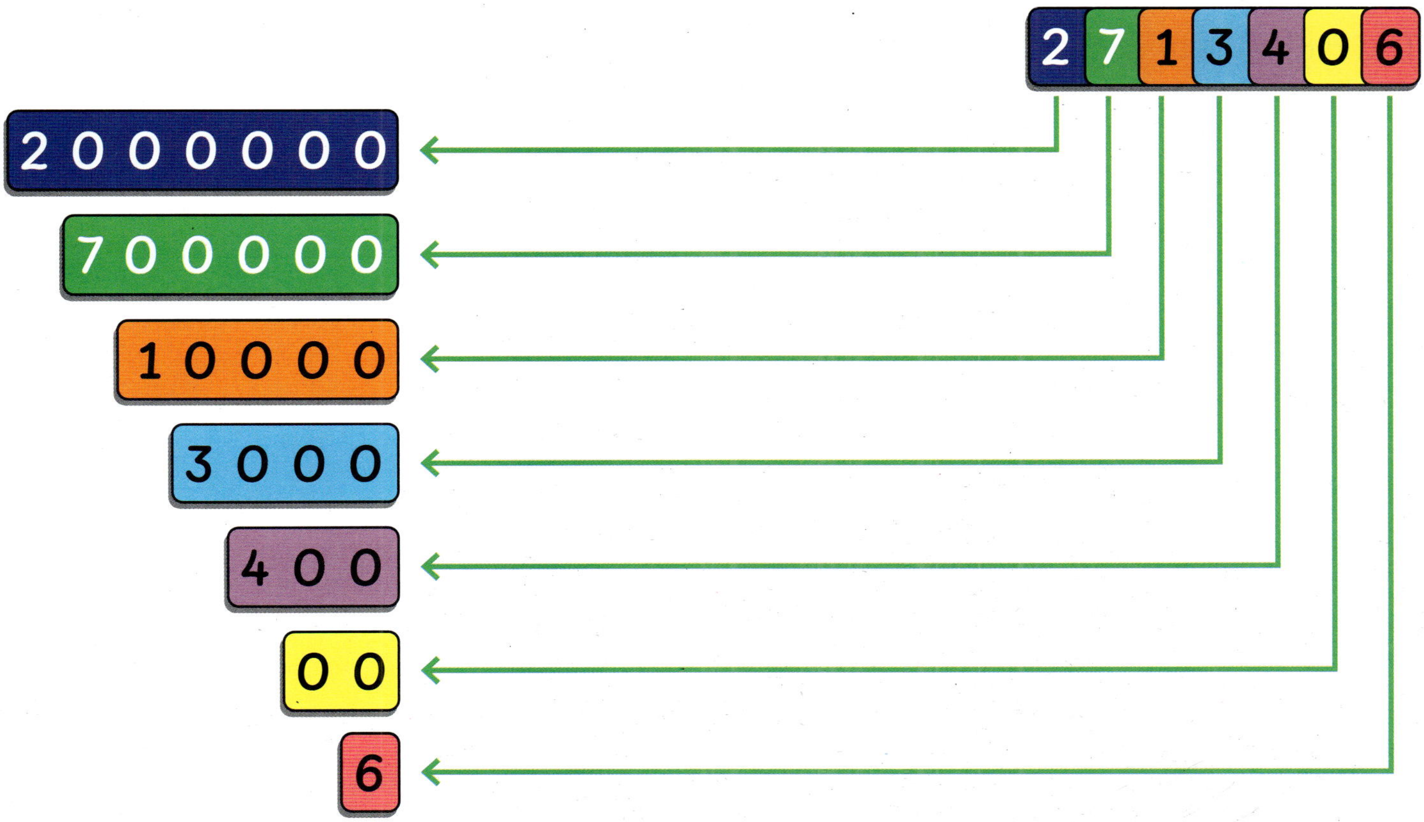

The digit 2 is in the millions place.
It has a value of 2 millions or 2 000 000.

The digit 7 is in the hundred thousands place.
It has a value of 7 hundred thousands or 700 000.

The digit 1 is in the ten thousands place.
It has a value of 10 thousands or 10 000.

The digit 3 is in the thousands place.
It has a value of 3 thousand or 3000.

The digit 4 is in the hundreds place.
It has a value of 4 hundreds or 400.

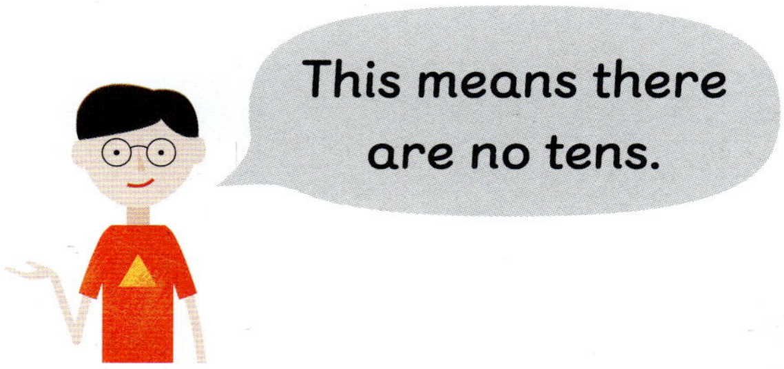

The digit 0 is in the tens place.
It has a value of 0.

The digit 6 is in the ones place.
It has a value of 6 ones or 6.

2 713 406 = 2 000 000 + 700 000 + 10 000 + 3 000 + 400 + 0 + 6

In words, we write 2 713 406 as two million, seven hundred and thirteen thousand, four hundred and six.

Guided Practice

The table shows the approximate population of some countries.

	Costa Rica	Namibia	Austria	Guinea-Bissau
Population in numerals	5 094 118		9 006 398	
Population in words		Two million, five hundred and forty thousand, nine hundred and five		One million, nine hundred and sixty-eight thousand and one

1 Complete the table.

2 In 5 094 118, the digit ___ is in the millions place and its value is ___.

3 In 9 006 398, the digit 6 is in the ___ place and its value is ___.

4 What value does the digit 9 have in each number?

5 Write each number in expanded form.

In expanded form, we show a number as the sum of the values of its digits.

$9\,006\,398 = 9\,000\,000 + 6000 + \square + \square + 8$

Complete Worksheet **1**: pages **1–2**

Comparing Numbers to 10 Million

Lesson 2

Explore

Amira rolls a  seven times and gets the following digits: 0, 3, 9, 5, 4, 8, 2.
What is the greatest 7-digit number she can make?
What is the smallest 7-digit number she can make?

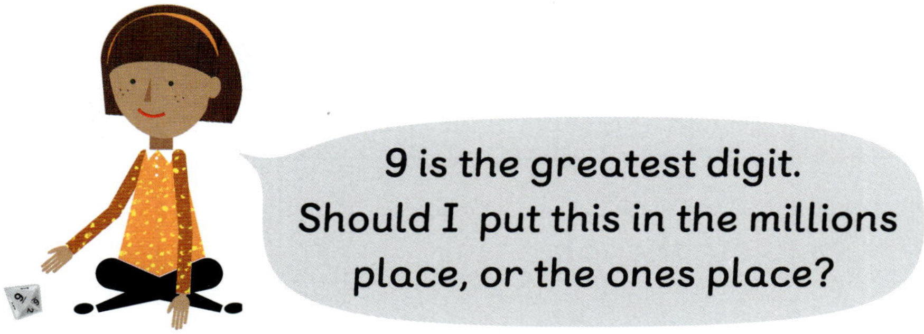

Master

1 Amira thinks that 9 085 432 is the greatest 7-digit number she can make, and that 0 234 589 is the smallest.
Is she correct?

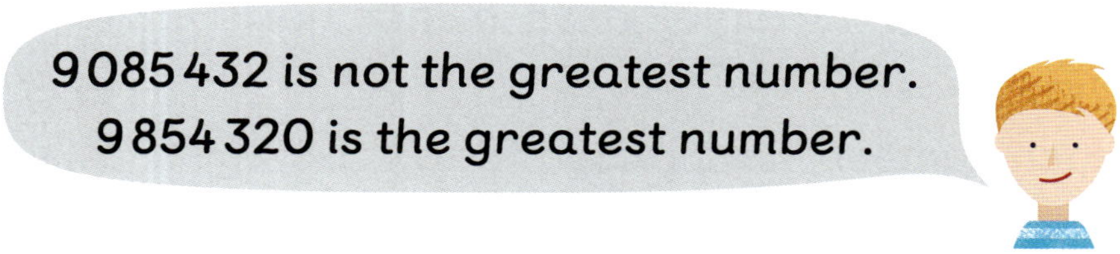

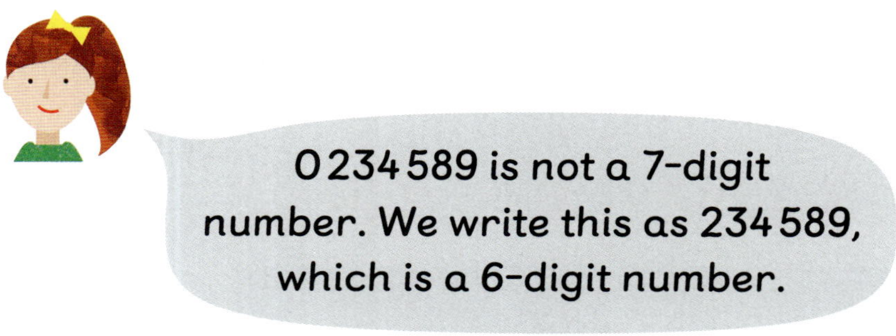

2 Sam thinks that 9 854 320 is the greatest 7-digit number he can make, and that 2 034 589 is the smallest.
Is he correct?

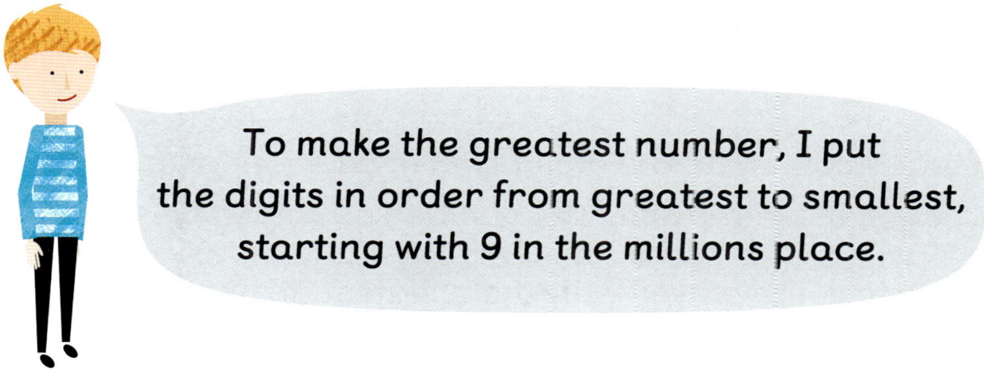

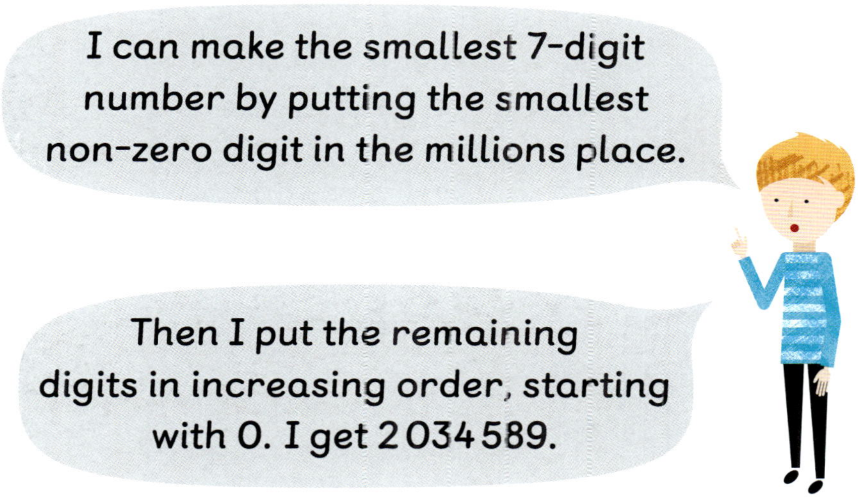

3 What are the possible missing digits?

8 ▨ 4 3 ▨ 1 ▨ > 8 5 6 3 ▨ ▨ 4

Activity Time

Play a game (2 pupils)

You will need:

 2 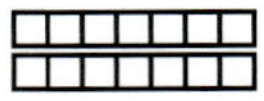

1. Take turns to roll  .
2. After each roll, write the digit down in your opponent's grid. Try to make the smallest number possible!
3. The winner is the player with the greatest number on the grid after each player has taken seven turns.

Guided Practice

Use > or < to fill in the blanks.

1 6 000 000 ____ 5 999 999

2 1 234 567 ____ 1 324 675

3 5 656 556 ____ 5 656 656

4 2 435 909 ____ 2 434 999

5 8 998 899 ____ 8 989 899

Complete Worksheet **2**: pages **3–4**

Comparing and Ordering Numbers to 10 Million

Lesson 3

Explore

Jacob is comparing the total area of different countries.

Country		Total area in km^2
Argentina		2 780 400
Denmark		2 220 093
Greenland		2 166 086
India		3 287 263
Kazakhstan		2 724 900

Which country in this table is the largest?
Which country in this table is the smallest?

Master

1 Compare the digits in the millions place.

	millions	hundred thousands	ten thousands	thousands	hundreds	tens	ones
	2	7	8	0	4	0	0
	2	2	2	0	0	9	3
	2	1	6	6	0	8	6
	3	2	8	7	2	6	3
	2	7	2	4	9	0	0

3 287 263 is the greatest number.
India is the largest country.

3 millions is greater than 2 millions.

2 More than one number has the same digit in the millions place.
Compare the digits in the hundred thousands place.

	millions	hundred thousands	ten thousands	thousands	hundreds	tens	ones
	2	7	8	0	4	0	0
	2	2	2	0	0	9	3
	2	1	6	6	0	8	6
	3	2	8	7	2	6	3
	2	7	2	4	9	0	0

2 166 086 is the smallest number.
Greenland is the smallest country.

When the digits in the greatest place value are the same, compare the digits in the next greatest place value.

3 Arrange the numbers in order from smallest to greatest.

	millions	hundred thousands	ten thousands	thousands	hundreds	tens	ones
	2	7	8	0	4	0	0
	2	2	2	0	0	9	3
	2	1	6	6	0	8	6
	3	2	8	7	2	6	3
	2	7	2	4	9	0	0

2 166 086, 2 220 093, 2 724 900, 2 780 400, 3 287 263

smallest → greatest

Guided Practice

1 Arrange the numbers in order from greatest to smallest.

5 099 167, 6 179 053, 6 305 880, 6 158 413, 6 304 997

2 Use > or < to fill in the blanks.

(a) 1 406 112 ___ 1 491 046 ___ 1 491 400

(b) 3 954 136 ___ 3 945 099 ___ 3 944 499

Complete Worksheet **3**: pages **5–6**

Rounding Numbers

Lesson 4

Explore

The table shows the number of passengers flying on some popular air routes during one year.
Can you estimate how many people flew on each route?

Route	Number of passengers
Sydney to Melbourne, Australia	9 245 392
Cancun to Mexico City, Mexico	4 885 602
Jakarta to Denpasar, Indonesia	5 535 108

Master

1 Find the approximate number of passengers flying on the Sydney–Melbourne air route.

Round 9 245 392 to the nearest 1 000 000.

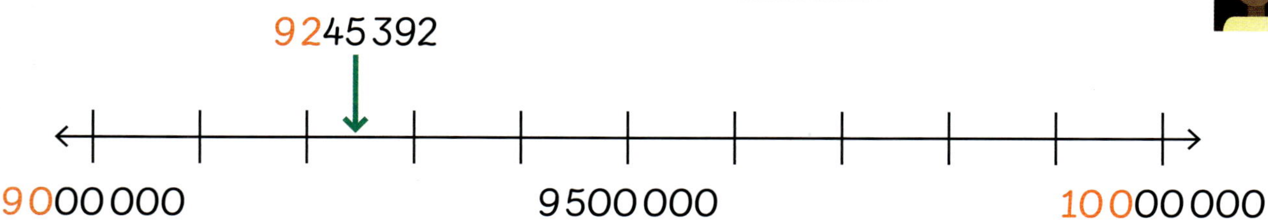

9 245 392 is closer to 9 000 000 than 10 000 000.

9 245 392 is approximately equal to 9 000 000.

9 245 392 ≈ 9 000 000 (rounded to the nearest 1 000 000)

Approximately 9 000 000 people flew on the Sydney–Melbourne air route.

2 Find the approximate number of passengers flying on the Cancun–Mexico City air route.

Round 4 885 602 to the nearest 1 000 000.

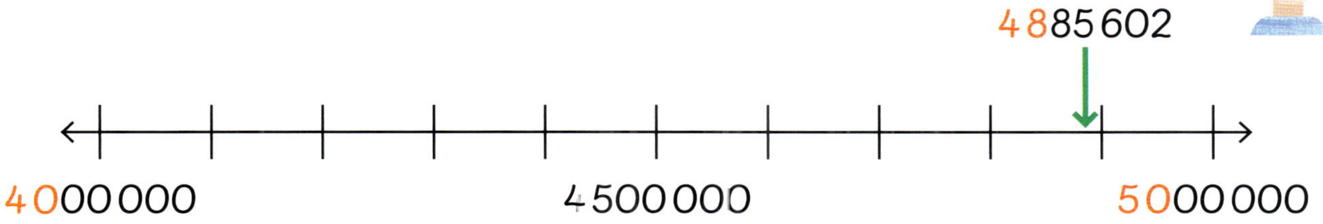

4 885 602 is closer to 5 000 000 than 4 000 000.

4 885 602 is approximately equal to 5 000 000.

4 885 602 ≈ 5 000 000 (rounded to the nearest 1 000 000)

Approximately 5 000 000 people flew on the Cancun–Mexico City air route.

3 Find the approximate number of passengers flying on the Jakarta–Denpasar air route.

Is 5 535 108 closer to 5 000 000 or 6 000 000?

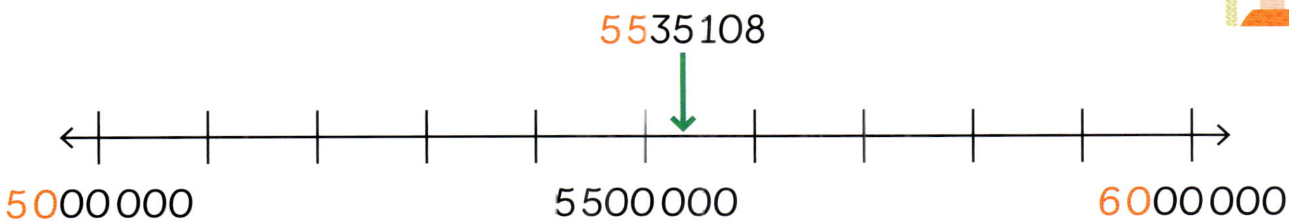

5 535 108 is approximately equal to 6 000 000.

5 535 108 ≈ 6 000 000 (rounded to the nearest 1 000 000)

Approximately 6 000 000 people flew on the Jakarta–Denpasar air route.

4 Round 9 500 000 to the nearest 1 000 000.

9 500 000 is exactly halfway between 9 000 000 and 10 000 000. So, we round up.

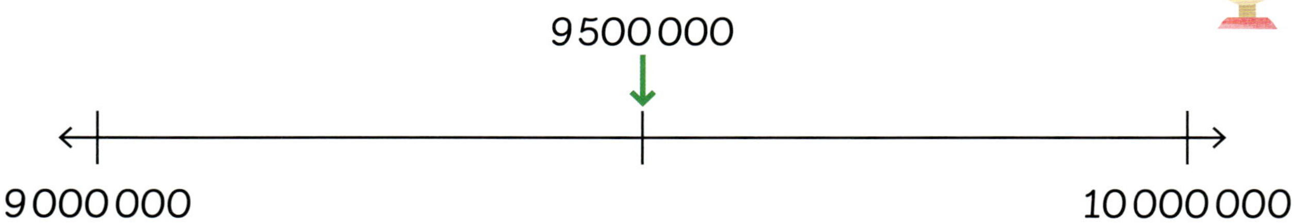

9 500 000 is approximately equal to 10 000 000.

9 500 000 ≈ ______ (rounded to the nearest 1 000 000)

5 New York JFK–Los Angeles is the busiest air route in the US. This route had 4 000 000 passengers, rounded to the nearest 1 000 000. What is the smallest and greatest possible number of passengers that could have flown this route?

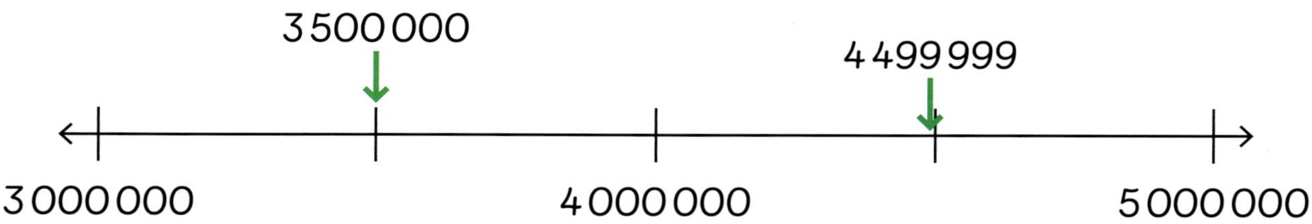

The smallest possible number is 3 500 000.
The greatest possible number is 4 499 999.
Both of these numbers round to 4 000 000 when rounded to the nearest 1 000 000.

The actual recorded number was 3 971 922 passengers.

Guided Practice

1 Round these numbers to the nearest 1 000 000.

(a) 7 375 890

(b) 1 854 735

(c) 2 526 892

(d) 4 500 000

2 Which of these numbers give the same number when rounded to the nearest 1 000 000?

3 495 077 4 388 432 4 500 976 3 612 100

3 Circle the number that is greater when rounded to the nearest 1 000 000.

(a) 7 049 462 6 487 940

(b) 5 477 087 5 747 088

(c) 8 893 560 8 393 991

4 The number of passengers flying the Madrid–Barcelona air route was 2 000 000 when rounded to the nearest 1 000 000.

(a) Find the smallest possible number of passengers on this route.

(b) Find the greatest possible number of passengers on this route.

Complete Worksheet **4**: pages **7–8**

Rounding Numbers

Lesson 5

Explore

Elliott is writing a presentation about India.

Area

Total area in km^2	3 287 263
Land area in km^2	2 973 190
Area of water bodies in km^2	314 073

Master

1 Elliott could round to the nearest 1 000 000.

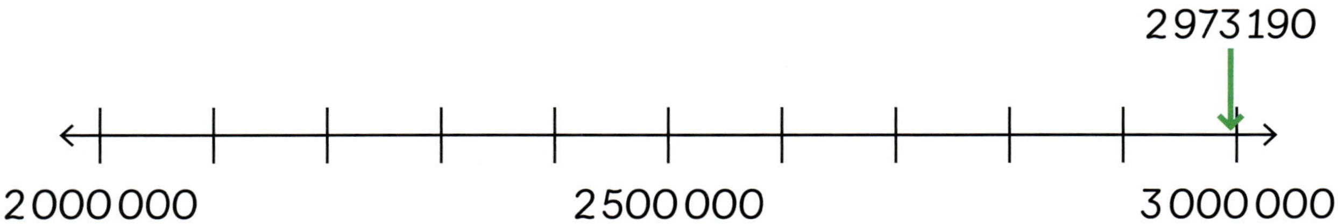

2 973 190 ≈ 3 000 000 (rounded to the nearest 1 000 000)
The land area of India is approximately 3 000 000 km^2.

2 Elliott could round to the nearest 100 000.

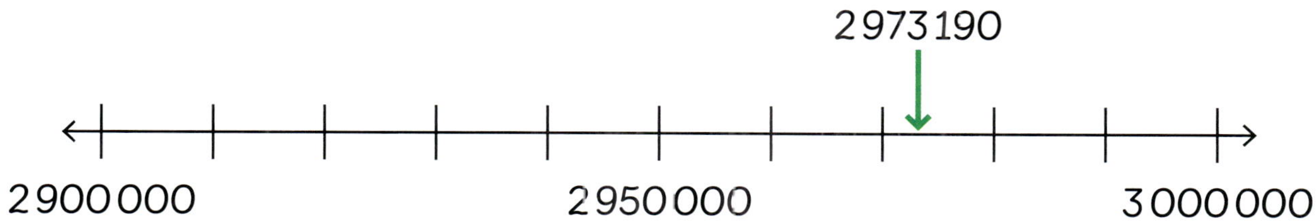

2 973 190 ≈ 3 000 000 (rounded to the nearest 100 000)
The land area of India is approximately 3 000 000 km^2.

3 Elliott could round to the nearest 10 000.

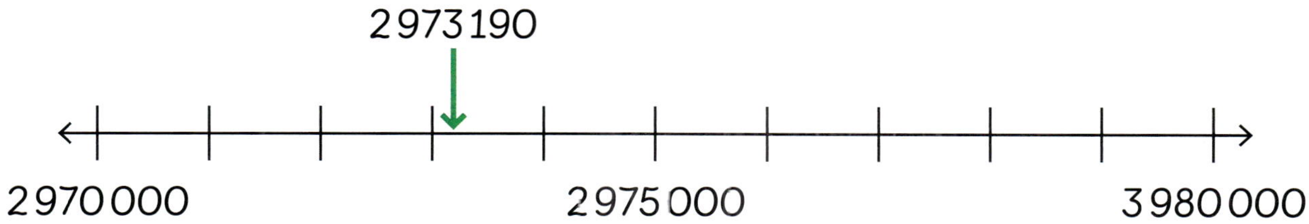

2 973 190 ≈ 2 970 000 (rounded to the nearest 10 000)
The land area of India is approximately 2 970 000 km^2.

4 Elliott could round to the nearest 1000.

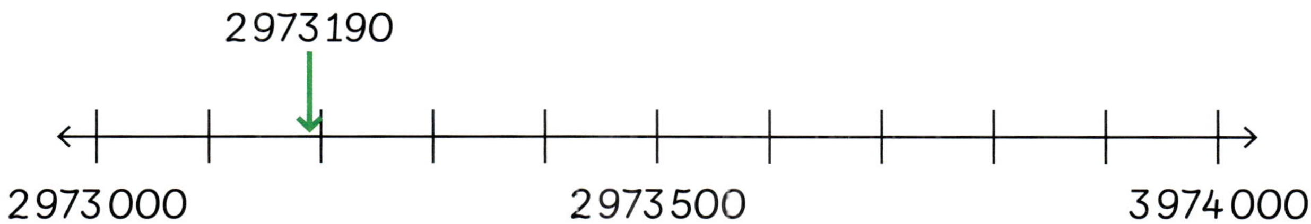

2 973 190 ≈ 2 973 000 (rounded to the nearest 1000)
The land area of India is approximately 2 973 000 km^2.

5 Elliott could round to the nearest 100.

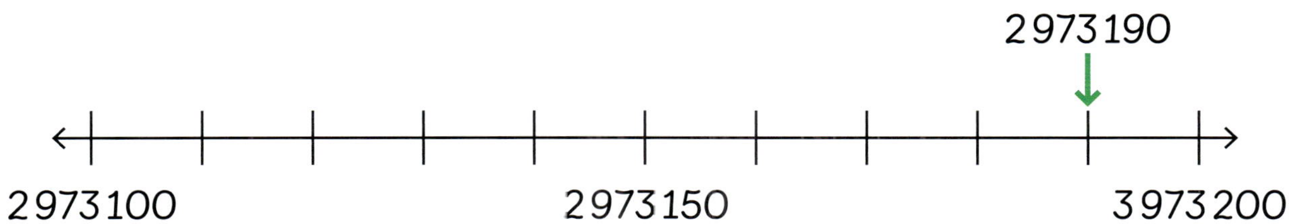

2 973 190 ≈ 2 973 200 (rounded to the nearest 100)
The land mass of India is approximately 2 973 200 km^2.

What number do you think Elliott should round to? Why?

Guided Practice

1 (a) Round 438 000 to:

(i) the nearest 10 000

(ii) the nearest 100 000

(b) Round 8 200 734 to:

(i) the nearest 1000

(ii) the nearest 100 000

(iii) the nearest 1 000 000

2 How can we round these numbers?

(a) 20 964 to the nearest 10 000

(b) 3408 to the nearest 1000

(c) 150 to the nearest 100

3 A number is 6 100 000 when rounded to the nearest 100 000.
What are the greatest and smallest possible values of this number?

Complete Worksheet **5**: pages **9–10**

Mind Challenge

This is the number system that the Ancient Egyptians used.

1	2	3	4	5	6	7	8	9	10

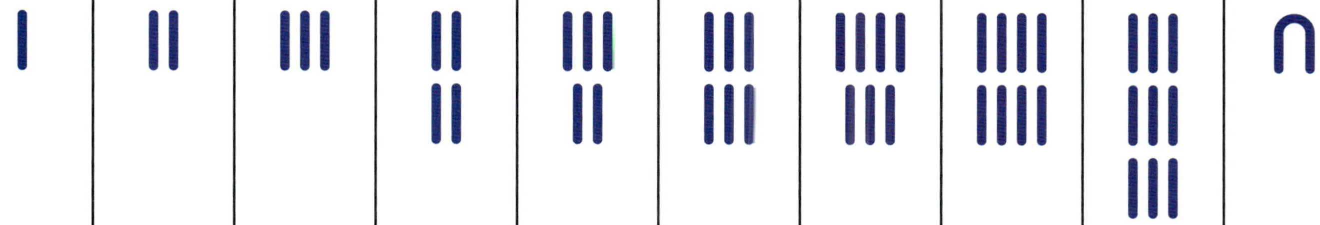

100	1000	10 000	100 000	1 000 000

Unlike the place-value system we use, this is an additive system.
Here is an example.

= 2 1 2 3 7

How would you show 2 345 678 using this system?

My Maths Journal

Is it always appropriate to round numbers?
Do you think it makes sense to round these numbers?

- 142 children having school dinners, rounded to the nearest 100.
- A population of 2 387 562 people, rounded to the nearest 100 000.
- A salary of £34 321, rounded to the nearest 10 000.

Explain your answer.

Review

I can...

- ☐ read and write numbers to 10 million.
- ☐ tell the place value of a digit in a number.
- ☐ compare and order numbers up to 10 million.
- ☐ round numbers to the nearest 10, 100, 1000, 10 000, 100 000 and 1 000 000.

Chapter 2

Four Operations on Whole Numbers

Using Mixed Operations

Lesson 1

Explore

Lulu, Sam and Holly are trying to solve this problem.

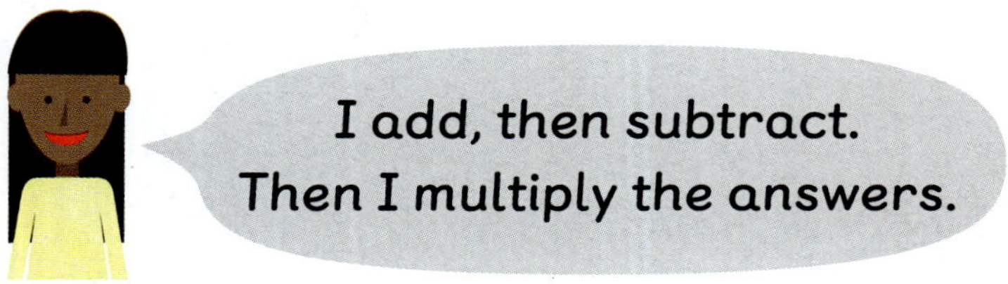

Who is correct?

Master

1 First, multiply or divide, working from left to right. Then, add or subtract, working from left to right.

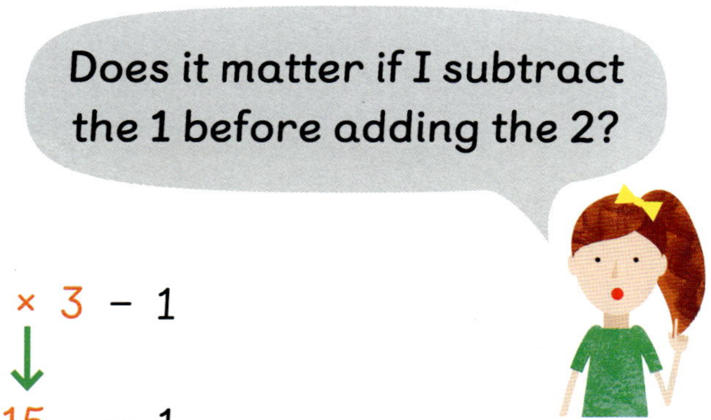

$$2 + 5 \times 3 - 1$$
$$= 2 + 15 - 1$$
$$= 17 - 1$$
$$= 16$$

Holly is correct.

2 Calculate $12 + 30 \div 5 \times 4$.

$$12 + 30 \div 5 \times 4$$
$$= 12 + 6 \times 4$$
$$= 12 + 24$$
$$= 36$$

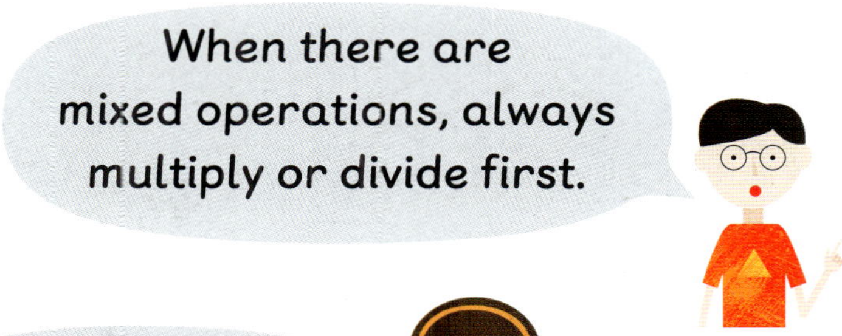

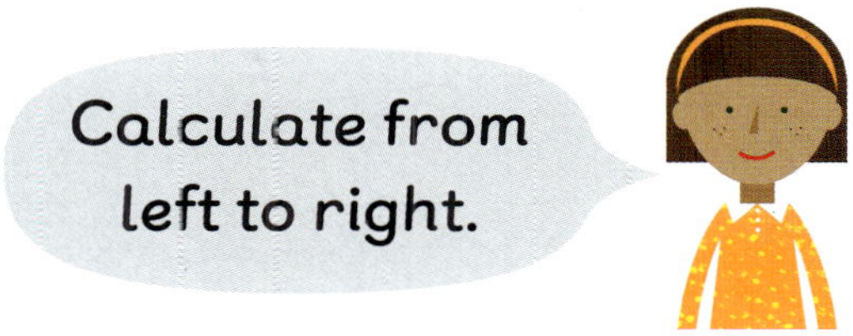

Guided Practice

Calculate.

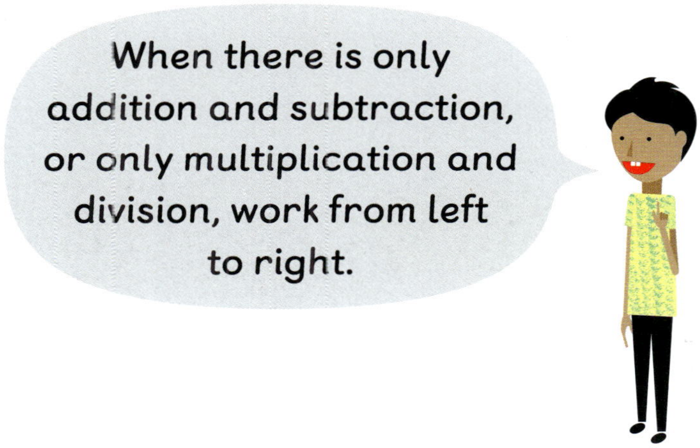

1

(a) $25 + 7 - 13 =$ ☐

(b) $51 - 33 + 26 =$ ☐

(c) $72 \div 8 \times 9 =$ ☐

(d) $7 \times 6 \div 2 =$ ☐

2

(a) $43 - 31 \times 4 =$ ☐

(b) $29 + 64 \div 8 =$ ☐

(c) $15 + 3 \times 9 - 27 =$ ☐

(d) $30 - 42 \div 7 + 18 =$ ☐

(e) $95 - 36 \div 4 \times 8 =$ ☐

(f) $41 + 8 \times 6 \div 4 =$ ☐

Complete Worksheet **1**: pages **15–16**

Order of Operations

Lesson 2

Explore

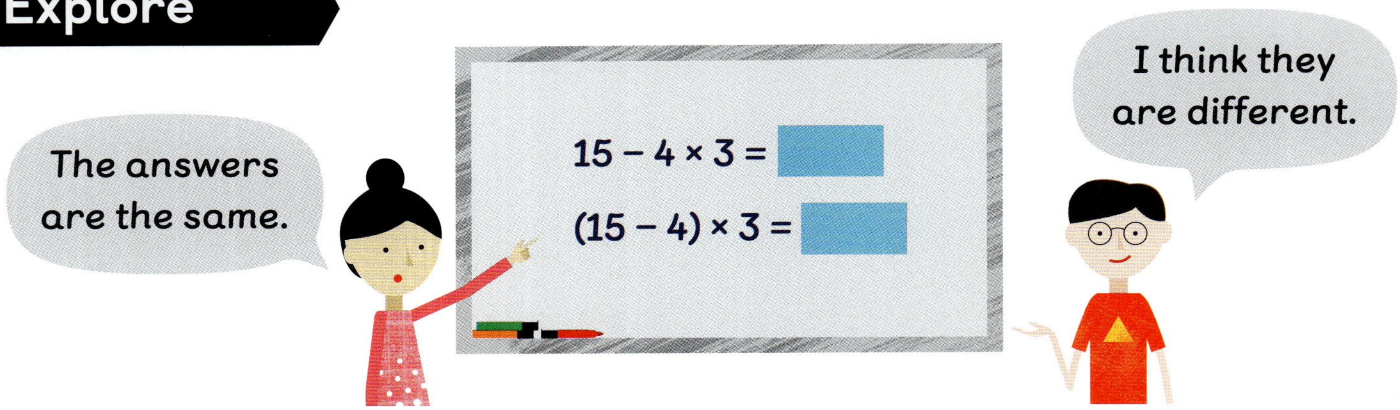

Who is correct?

Master

1 First, carry out all the operations in ( ).
Next, perform all the multiplication and division.
Then, calculate all the addition and subtraction.

$15 - 4 \times 3 = 15 - 12$
$= 3$

$(15 - 4) \times 3 = 11 \times 3$
$= 33$

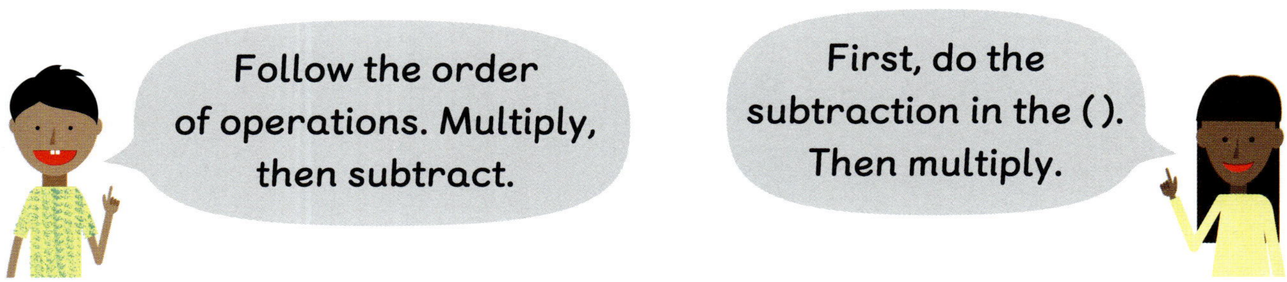

Charles is correct.

Guided Practice

1 Calculate.

(a) $(1 + 3) \times 5 - 7 =$

(b) $1 + (3 \times 5) - 7 =$

(c) $(1 + 3) \times (7 - 5) =$

2 Calculate.

(a) $5 \times (5 \times 4) =$

(b) $90 \div (63 \div 7) =$

(c) $(12 + 18) \div 6 =$

(d) $(35 - 28) \times 4 =$

(e) $16 \div 4 + (3 \times 2) =$

(f) $40 - (8 + 19) \div 3 =$

3 Insert ( ) where needed to make these calculations correct.

(a) $3 + 6 \div 2 - 1 = 9$

(b) $3 + 6 \div 2 - 1 = 5$

(c) $3 + 6 \div 2 + 1 = 5$

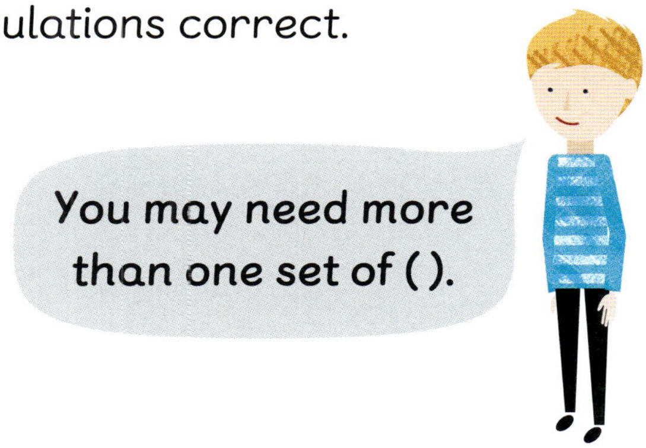

Complete Worksheet **2**: pages **17–18**

Multiplying by Tens

Lesson 3

Explore

How many grams of cereal are there in 20 boxes?

Master

1 $432 \times 10 = \square$

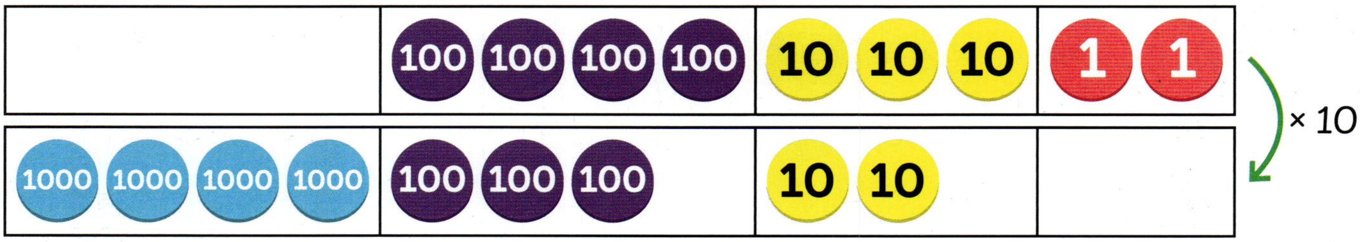

$432 \times 10 = 4320$

2 $432 \text{ g} \times 20 = \square$

Method 1

th	h	t	o
	4	3	2
4	3	2	0

× 10

$432 \times 10 = 4320$

$432 \times 20 = 4320 \times 2$

$= 8640$

Method 2

$432 \times 20 = 432 \times 2 \times 10$
$= 864 \times 10$
$= 8640$

There are 8640 g of cereal in 20 boxes.

3 Emma buys a family-size box that contains 1432 g of cereal.
How many grams of cereal are there in 20 family-size boxes?

Method 1

t th	th	h	t	o
	1	4	3	2
1	4	3	2	0

× 10

$1432 \times 10 = 14\,320$
$1432 \times 20 = 14\,320 \times 2$
$= 28\,640$

Method 2

$1432 \times 20 = 20\,000 + 8640$
$= 28\,640$

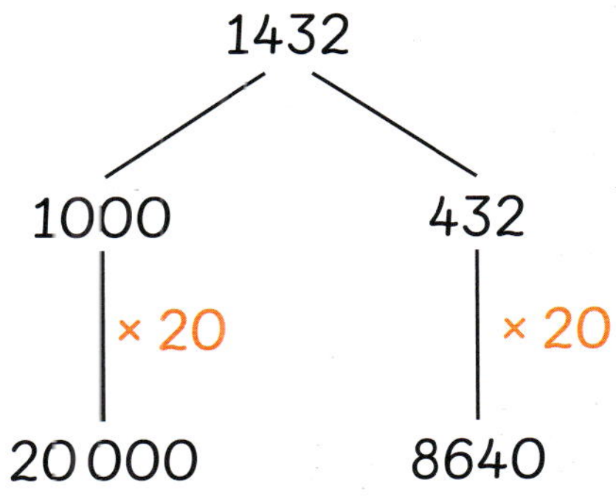

Method 3

$1432 \times 20 = 1432 \times 2 \times 10$
$= 2864 \times 10$
$= 28\,640$

There are 28 640 g of cereal in 20 family-size boxes.

Guided Practice

1 Multiply.

(a) 123 × 10 =

(b) 123 × 20 =

(c) 123 × 30 =

(d) 1123 × 30 =

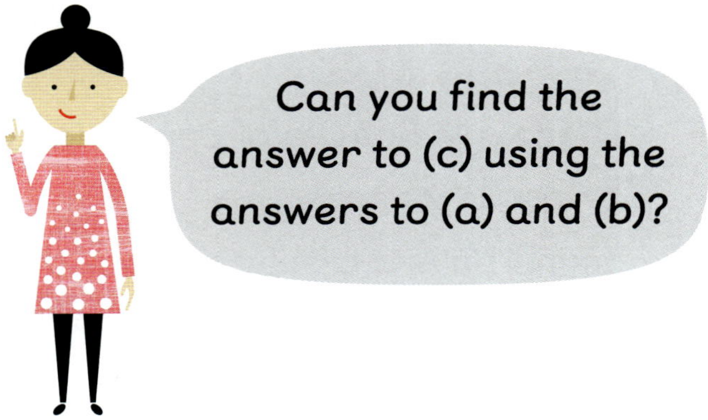

2 Find the product.

(a) 1212 and 40 =

(b) 2424 and 20 =

(c) 2414 × 50 =

(d) 1212 × 100 =

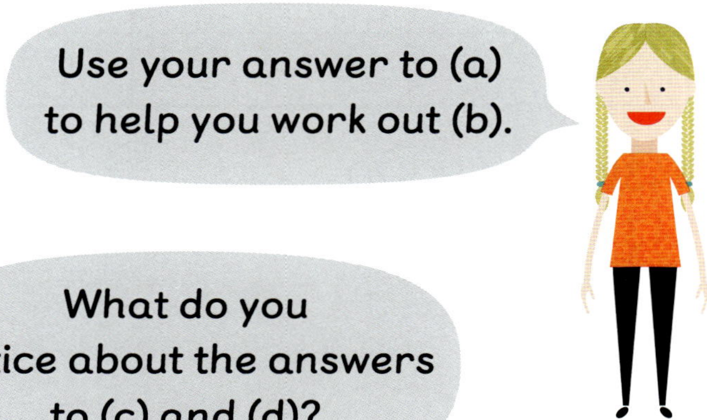

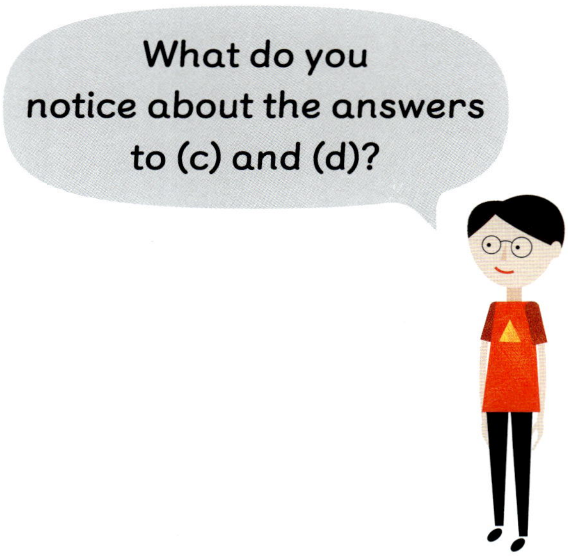

Complete Worksheet **3**: pages **19–20**

Multiplying a 3-Digit Number by a 2-Digit Number

Lesson 4

Explore

Mr Nightingale wants to buy 21 tablets for his class.
He has £3000 to spend. Does he have enough money?

Master

1 142 × 20 =

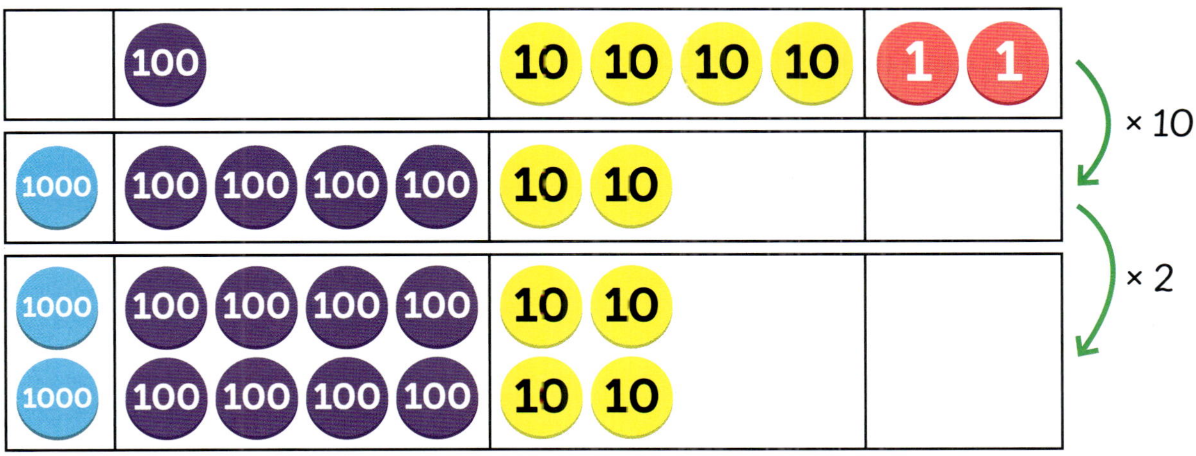

142 × 20 = 2840

2 $142 \times 21 =$ ___

th	h	t	o
	1	4	2
1	4	2	0

× 10

$142 \times 10 = 1420$

$142 \times 20 = 1420 \times 2$
$= 2840$

$142 \times 21 = 2840 + 142$
$= 2982$

3 $£142 \times 21 =$ ___

		1	4	2	
×			2	1	
		1	4	2	→ $142 \times 1 = 142$
+	2	8	4	0	→ $142 \times 20 = 2840$
	2	9	8	2	→ $142 \times 21 = 2982$

He has enough money to buy 21 tablets.

Guided Practice

1 Multiply.

(a) 123 × 20 = ___

(b) 123 × 4 = ___

(c) 123 × 24 = ___

2 Multiply.

(a) 32 × 150 = ___

(c) 32 × 30 = ___

(b) 320 × 15 = ___

(d) 16 × 60 = ___

3 Fill in the missing digits.

(a)

		1	2	4
×			1	___
		2	4	___
+	___	2	4	0
	1	4	___	8

(b)

		3	2	1
×			___	3
		___	6	3
+	3	2	___	0
	___	1	7	3

Complete Worksheet **4**: pages **21–22**

Multiplying by a 2-Digit Number

Lesson 5

Explore

A board game contains 310 question cards. How many question cards are there in 23 sets of the game?

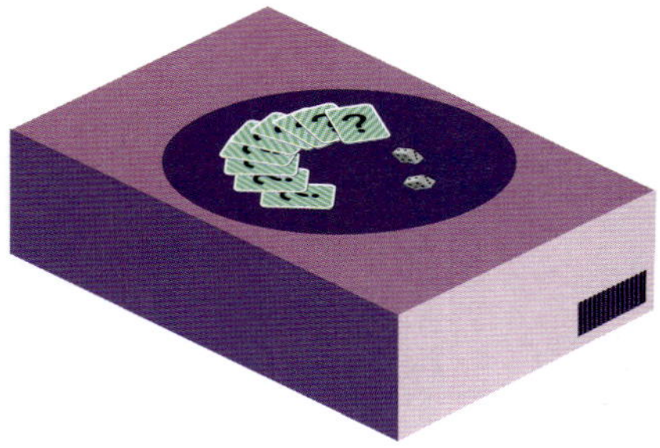

Master

1 310 × 20 = ____

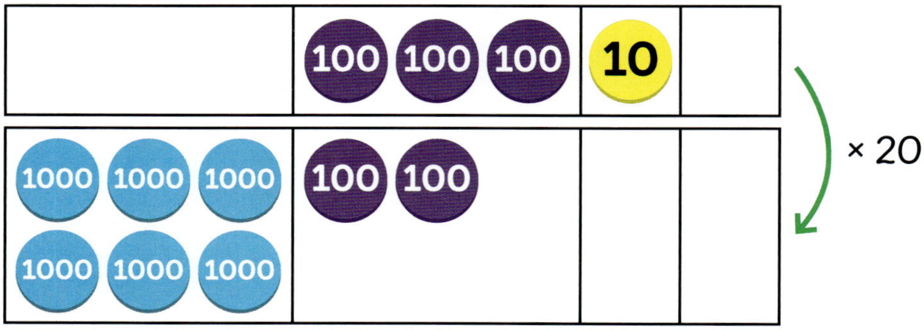

310 × 20 = 6200

2 310 × 23 = ____

th	h	t	o
	3	1	0
3	1	0	0

× 10

310 × 10 = 3100
310 × 20 = 6200
310 × 23 = 6200 + 930
= 7130

3 310 × 23 = ____

		3	1	0	
×			2	3	
		9	3	0	→ 310 × 3 = 930
+	6	2	0	0	→ 310 × 20 = 6200
	7	1	3	0	→ 310 × 23 = 7130

There are 7130 question cards in 23 sets of the game.

4 Find the product of 1310 and 23.

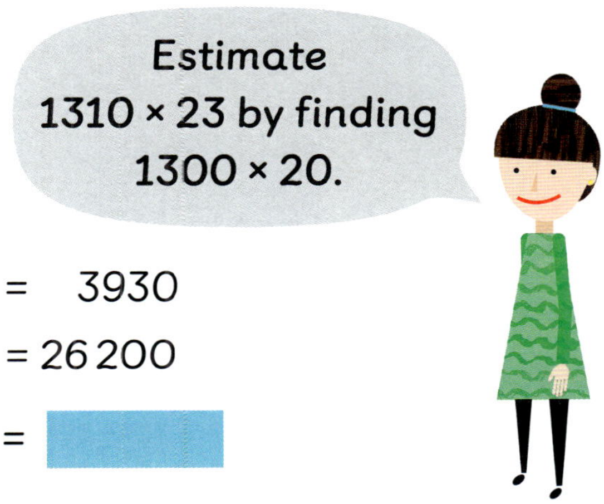

		1	3	1	0	
×				2	3	
		3	9	3	0	→ 1310 × 3 = 3930
+	2	6	2	0	0	→ 1310 × 20 = 26 200
						→ 1310 × 23 = ____

Guided Practice

Multiply.

1
(a) 322 × 10 = ____
(b) 322 × 20 = ____
(c) 322 × 30 = ____
(d) 322 × 3 = ____
(e) 322 × 13 = ____

2

(a) $345 \times 10 =$ ☐

(b) $345 \times 20 =$ ☐

(c) $345 \times 2 =$ ☐

(d) $345 \times 12 =$ ☐

(e) $1345 \times 22 =$ ☐

3

(a)

$$\begin{array}{r} 2\ 1\ 3 \\ \times \quad 2\ 3 \\ \hline \square \\ +\ \square \\ \hline \square \\ \hline \end{array}$$

(b)

$$\begin{array}{r} 1\ 2\ 1\ 3 \\ \times \quad 2\ 3 \\ \hline \square \\ +\ \square \\ \hline \square \\ \hline \end{array}$$

(c)

$$\begin{array}{r} 3\ 2\ 2 \\ \times \quad 3\ 3 \\ \hline \square \\ +\ \square \\ \hline \square \\ \hline \end{array}$$

(d)

$$\begin{array}{r} 1\ 3\ 2\ 2 \\ \times \quad 3\ 3 \\ \hline \square \\ +\ \square \\ \hline \square \\ \hline \end{array}$$

Complete Worksheet **5**: pages **23–24**

Multiplying a 3-Digit Number by a 2-Digit Number

Lesson 6

Explore

A zookeeper prepares 825 kg of food for the zoo's elephants each week.
What is the total mass of food the zookeeper prepares for the elephants over 16 weeks?
What is the total mass of food the zookeeper prepares for the elephants over 32 weeks?

Master

1 825 kg × 16 = ☐

			^{1}8	^{3}2	5
×				1	6
		4	9	5	0
+		8	2	5	0
	1	3	2	0	0

→ 825 × 6 = 4950

→ 825 × 10 = 8250

→ 825 × 16 = 13 200

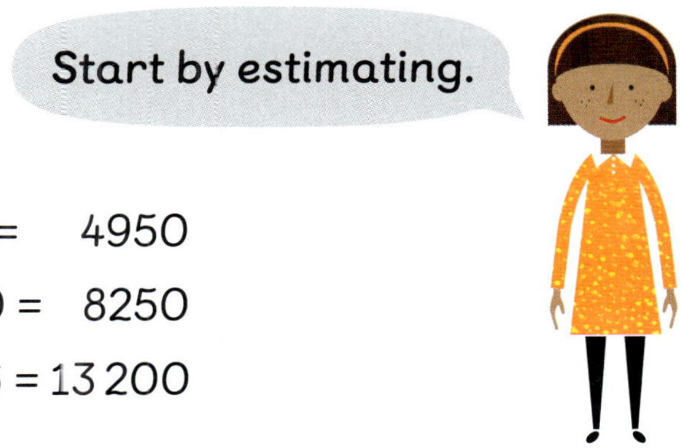

The zookeeper prepares 13 200 kg of food for the elephants over 16 weeks.

2 825 kg × 32 = [ ]

			8	¹₁ 2	5
×				3	2
		1	6	5	0
+	2	4	7	5	0
	2	6	4	0	0

→ 825 × 2 = 1650

→ 825 × 30 = 24 750

→ 825 × 32 = 26 400

The zookeeper prepares 26 400 kg of food for the elephants over 32 weeks.

3 825 × 16 = 13 200

825 × 32 = [ ]

825 × 32 = 13 200 × 2

= 26 400

4 1825 × 32 = [ ]

1825 × 32 = 32 000 + 26 400

= 58 400

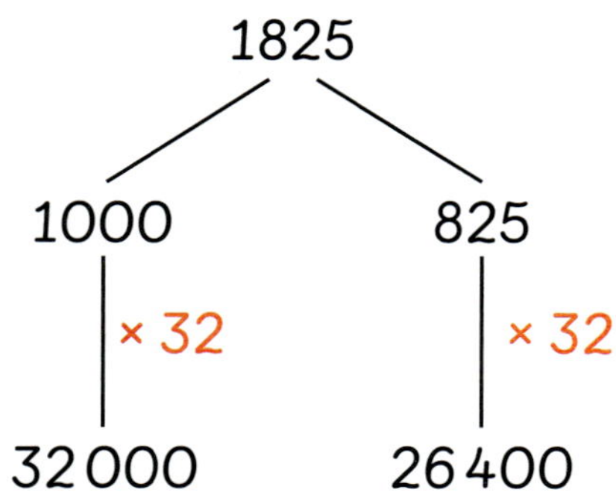

Guided Practice

1 Multiply.

$$\begin{array}{r} 241 \\ \times \quad 12 \\ \hline \square \\ + \quad \square \\ \hline \square \\ \hline \end{array}$$

→ 241 × ☐ = ☐

→ 241 × ☐ = ☐

→ 241 × ☐ = ☐

2 Multiply.

(a)
$$\begin{array}{r} 412 \\ \times \quad 16 \\ \hline \square \\ + \quad \square \\ \hline \square \\ \hline \end{array}$$

(b)
$$\begin{array}{r} 412 \\ \times \quad 32 \\ \hline \square \\ + \quad \square \\ \hline \square \\ \hline \end{array}$$

3 (a) Find the product of 534 and 24.

(b) Find the product of 1534 and 24.

Complete Worksheet **6**: pages **25–26**

Multiplying a 4-Digit Number by a 2-Digit Number

Lesson 7

Explore

A school buys 28 laptops.
Each laptop costs £1229.
How much does the school pay in total for the laptops?

Master

1 Estimate the product of 28 and 1229.

$28 \times 1229 \approx 30 \times 1000$
$= 30\,000$

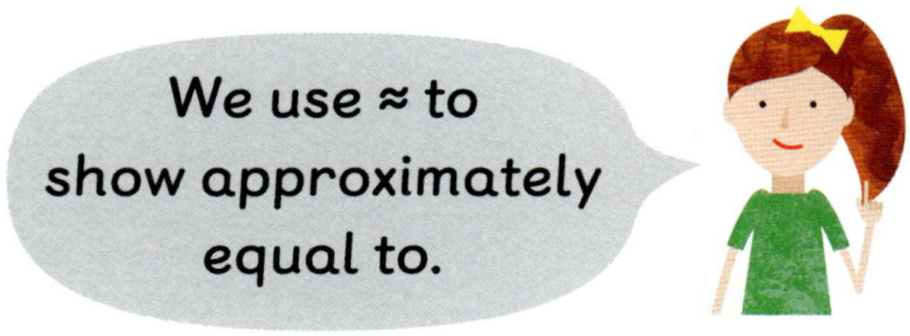

2 $28 \times 1229 =$ ▭

	^{1}1	^{2}2	^{7}2	9
×				8
	9	8	3	2

		1	2	^{1}2	9
×				2	0
	2	4	5	8	0

3 £1229 × 28 = [ ]

		^{1}1	^{2}2	$^{1}_{7}$2	9
×				2	8
		9	8	3	2
+	2	4	5	8	0
	3	4	4	1	2

→ 1229 × 8 = 9832
→ 1229 × 20 = 24 580
→ 1229 × 28 = 34 412

The school pays £34 412 in total for the laptops.

Guided Practice

1 Multiply.

(a)

	1	4	3	2
×			2	7
	[ ]			
+	[ ]			
	[ ]			

(b)

	3	5	1	7
×			3	4
	[ ]			
+	[ ]			
	[ ]			

2 Find the product of 2847 and 35.

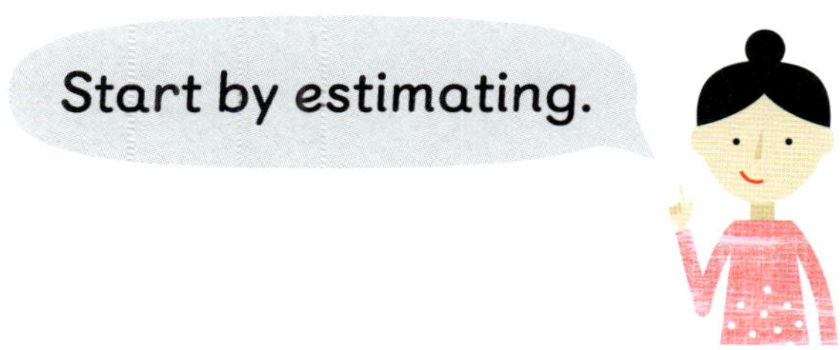

Complete Worksheet **7**: pages **27–28**

Multiplying by a 2-Digit Number

Lesson 8

Explore

Use these digit cards to make a 4-digit number and a 2-digit number to get a product that is a 5-digit number.
Each digit card can only be used once.

1 2 3 4 5 6

Master

1 Emma makes a guess with 6543 × 21.

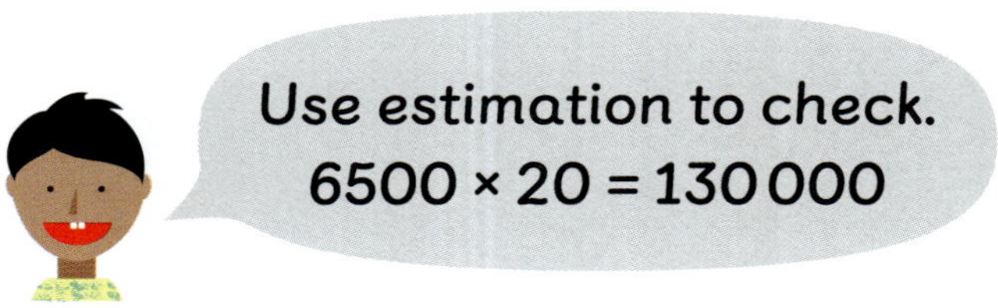

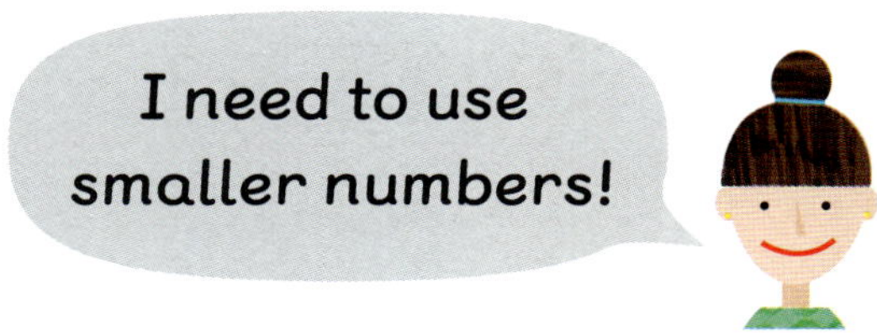

Emma makes another guess with 6543 × 12.

$$\begin{array}{r} {}^{1}6\,5\,4\,3 \\ \times \quad 1\,2 \\ \hline 1\,3\,0\,8\,6 \\ +\,6\,5\,4\,3\,0 \\ \hline 7\,8\,5\,1\,6 \\ \hline \end{array}$$

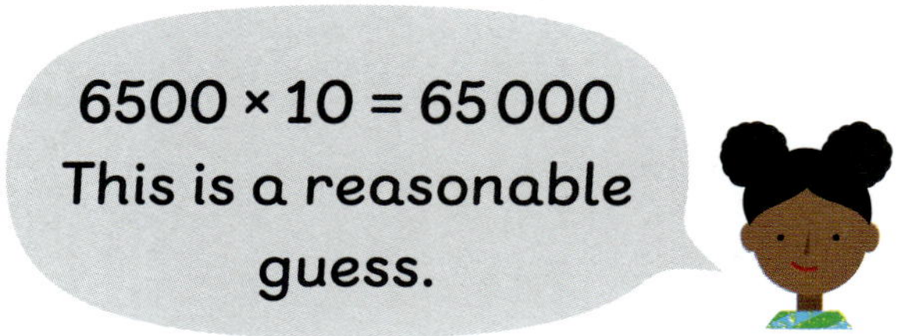

6543 × 12 = 78 516

2 Elliott wants to find the greatest possible product.

6000 × 10 = 60 000
6000 × 20 = 120 000
6000 × 15 = 90 000

I should use 15 as the 2-digit number.

He makes a guess with 6432 × 15.

		2	1	2	
		6	4	2	2
×				1	5
	3	2	1	6	0
+	6	4	3	2	0
	9	6	4	8	0

Do you think this is the greatest possible product?

3 Holly wants to find the smallest possible product.

First, I make the smallest 2-digit number, 12. So, the smallest 4-digit number I can make is 3456.

She makes a guess with 3456 × 12.

			1	1	
		3	4	5	6
×				1	2
		6	9	1	2
+	3	4	5	6	0
	4	1	4	7	2

Do you think this is the smallest possible product?

4 Charles has a different method to find the smallest possible product.

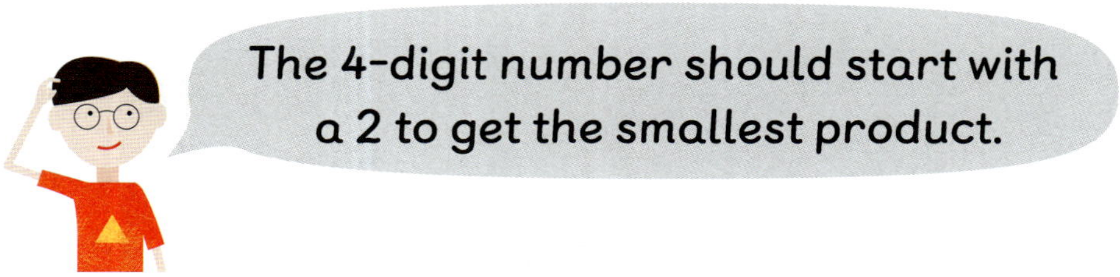

He makes a guess with 2456 × 13.

$$
\begin{array}{cccccc}
 & & {}^{1}2 & {}^{1}4 & {}^{1}4 & 6 \\
\times & & & & 1 & 3 \\
\hline
 & & 7 & 3 & 6 & 8 \\
+ & 2 & 4 & 5 & 6 & 0 \\
\hline
 & 3 & 1 & 9 & 2 & 8 \\
\hline
\end{array}
$$

Guided Practice

1 Using the digits 2, 3, 4, 5 and 6, make a 3-digit number and a 2-digit number that can be multiplied to give a 5-digit number. Each digit can only be used once.

(a) What numbers can you make to get the smallest possible product?

(b) What numbers can you make to get the greatest possible product?

2 Which expression will give a smaller product?
Estimate, then find the product to see if your guess is correct.

(a) 12 × 456   124 × 56

(b) 156 × 24   15 × 426

Complete Worksheet **8**: pages **29–30**

Dividing by a 2-Digit Number

Lesson 9

Explore

450 paintbrushes need to be packed into boxes of 15.
How many boxes are needed?

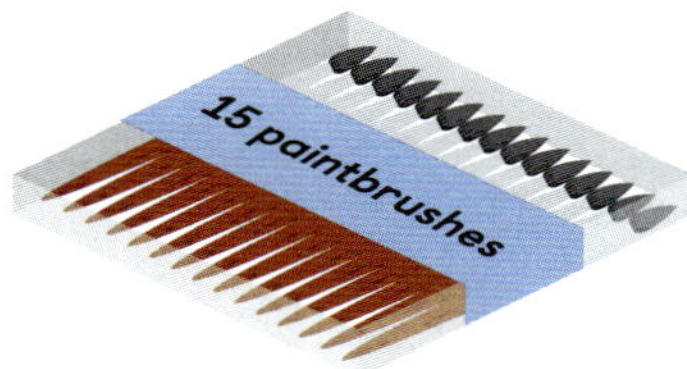

Master

1 $450 \div 15 = \square$

$45 \div 15 = 3$

$450 \div 15 = 30$

2 $450 \div 15 = \square$

45 tens ÷ 15 = 3 tens

$450 \div 15 = 30$

3 $450 \div 15 = \square$

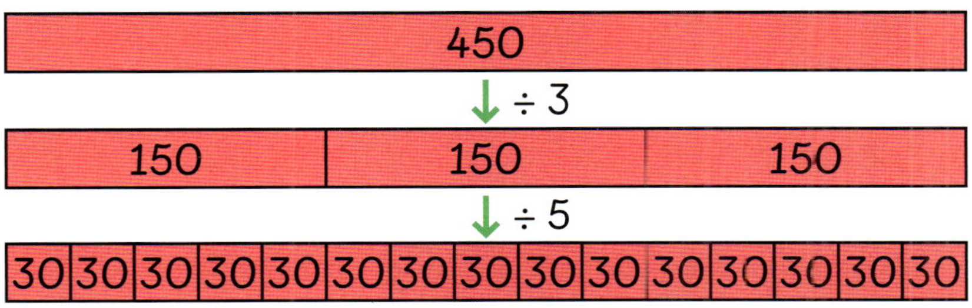

$450 \div 15 = 30$

30 boxes are needed to pack the paintbrushes.

Guided Practice

1 Divide.

(a) $450 \div 30 = $ ☐

(b) $4500 \div 30 = $ ☐

(c) $450 \div 90 = $ ☐

(d) $4500 \div 45 = $ ☐

(e) $4500 \div 15 = $ ☐

2 Divide.

(a) $360 \div 18 = $ ☐

(b) $720 \div 18 = $ ☐

(c) $3600 \div 18 = $ ☐

(d) $7200 \div 36 = $ ☐

3 920 tulips are arranged in rows of 46. How many rows are there?

4 240 children are divided into the following number of groups. How many children are there in each group?

(a) 12 (b) 20 (c) 15

Complete Worksheet **9**: pages **31–32**

Dividing by a 2-Digit Number

Lesson 10

Explore

Each day, 2448 eggs are collected from a farm.
They are arranged in trays that hold 24 eggs each.
How many trays are filled each day?

Master

1 $2448 \div 24 =$ ▭

```
       1 0 2
24 ) 2 4 4 8
   - 2 4          → 24 hundreds ÷ 24 = 1 hundred
         4 8
   -     4 8      → 48 ones ÷ 24 = 2 ones
           0
```

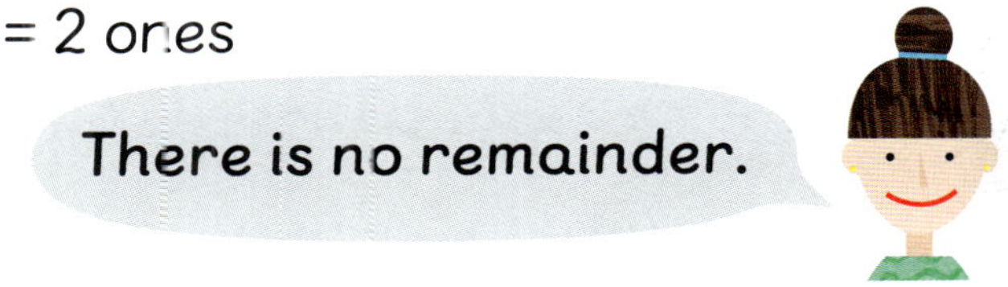

$2448 \div 24 = 102$
102 trays are filled each day.

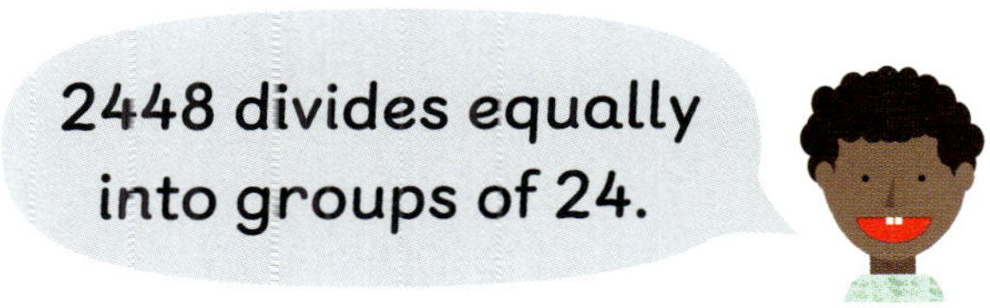

2

Can we use 1224 eggs to fill trays of 12?

How about 612 eggs in trays of 6?

$1224 \div 12 = \square$ $\quad$ $612 \div 6 = \square$

$$
\begin{array}{r}
102 \\
12\overline{)\,1224} \\
-\,12 \\
\hline
24 \\
-\,24 \\
\hline
0
\end{array}
$$

2448 eggs
24 eggs in a tray ⟶ 102 trays

1224 | 1224 eggs
12 eggs in a tray ⟶ 102 trays

612 | 612 eggs
6 eggs in a tray ⟶ 102 trays

We can use **repeated division** to find the answer to 2448 ÷ 24.

Guided Practice

1 Divide.

(a) $4900 \div 20 = \square$

(b) $2450 \div 10 = \square$

(c) $1225 \div 5 = \square$

2 1540 straws are packed equally into 14 boxes.
How many straws are packed in each box?

Complete Worksheet **10**: pages **33–34**

Dividing by a 2-Digit Number

Lesson 11

Explore

A machine can pack pens into boxes of 20 or 28 pens each. How many boxes can it pack from 4676 pens with each box being full?

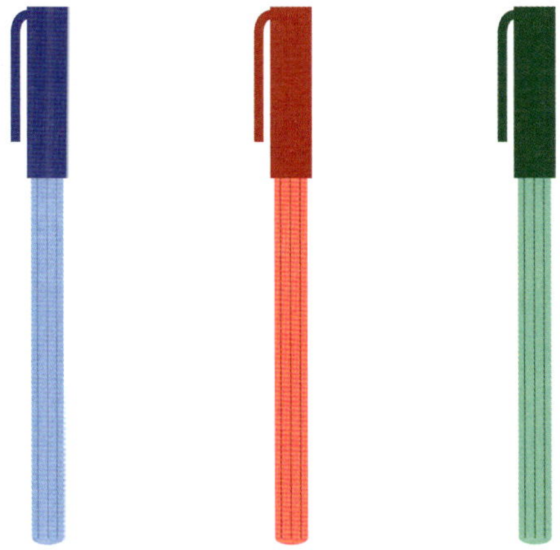

Master

1 4676 ÷ 20 = ▢

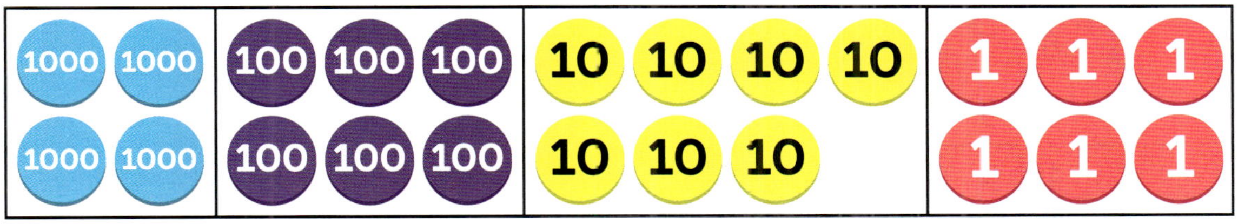

4676 → 4000 and 676

4000 ÷ 20 = 200 ← 4000
200 boxes

676 → 600 and 76

600 ÷ 20 = 30 ← 600
30 boxes

76 → 60 and 16

60 ÷ 20 = 3 ← 60
3 boxes

16

4676 ÷ 20 = 233 remainder 16

The machine can pack 233 boxes of 20 pens with 16 pens left over.

2 $4676 \div 28 =$ ▭

```
         1
28 ) 4 6 7 6
   - 2 8        →  28 hundreds ÷ 28 = 1 hundred
```

↓

```
         1 6
28 ) 4 6 7 6
   - 2 8        →  28 hundreds ÷ 28 = 1 hundred
     -----
     1 8 7
   - 1 6 8      →  168 tens ÷ 28 = 6 tens
     -----
```

↓

```
         1 6 7
28 ) 4 6 7 6
   - 2 8        →  28 hundreds ÷ 28 = 1 hundred
     -----
     1 8 7
   - 1 6 8      →  168 tens ÷ 28 = 6 tens
     -------
       1 9 6
   -   1 9 6    →  196 ones ÷ 28 = 7 ones
     -------
           0
```

$4676 \div 28 = 167$

The machine can pack 167 boxes of 28 pens with no pens left over.

3 Can the machine pack 364 pens into boxes of 28?

```
         1      →  280 ÷ 28 = 10
28 ) 3 6 ⁸4
```

↓

```
         1 3    →  84 ÷ 28 = 3
28 ) 3 6 4
```

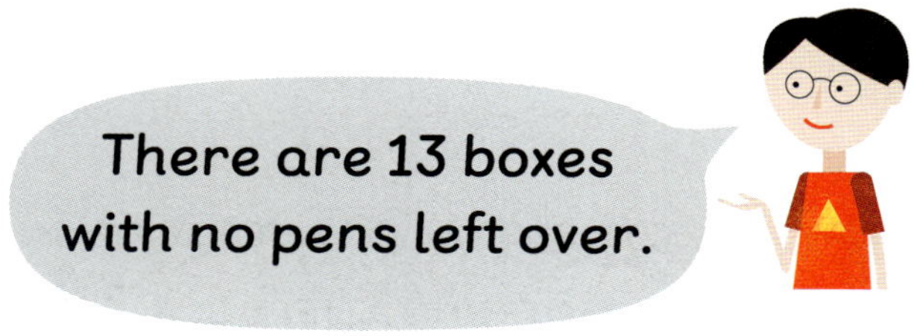

Guided Practice

1 Charles used three different methods to calculate 468 ÷ 36.

468 → 360 and 108

360 ÷ 36 = [ ] 108 ÷ 36 = [ ]

468 ÷ 36 = [ ]

```
       [ ][ ]
36 ) 4  6  8
   - 3  6
   ---------
     1  0  8
   - 1  0  8
   ---------
           0
```

```
      [ ][ ]
36 ) 4  6  ¹⁰8
```

Use these three methods to find:

(a) 462 ÷ 42

(b) 312 ÷ 26

(c) 396 ÷ 11

2 Find the value of 4386 ÷ 43.

3 3996 ÷ 18 = [ ]

```
        [ ][ ][ ]
18 ) 3  9  9  6
   - [ ][ ]
   ------------
        [ ][ ]
   -    [ ][ ]
   ------------
           [ ][ ]
   -       [ ][ ]
   ------------
              0
```

3996 → 3600 and [ ]

[ ] → [ ] and [ ]

Complete Worksheet 11: pages 35–36

Dividing by a 2-Digit Number with Remainder

Lesson 12

Explore

581 rulers are delivered to a school.
The rulers need to be divided equally between 18 classrooms.
Any remaining rulers will be kept in the school office.
How many rulers will each classroom receive?
How many rulers will be kept in the school office?

Master

1 $581 \div 18 =$

$$
\begin{array}{r|l}
 & \;\;3\;\;2 \text{ remainder } 5 \\
18 & \overline{\;\;5\;\;8\;\;1} \\
 & -\;5\;\;4\;\; \longrightarrow 3 \text{ tens} \times 18 = 54 \text{ tens} \\
 & \overline{\;\;\;\;\;\;4\;\;1} \\
 & -\;\;\;\;3\;\;6 \longrightarrow 2 \text{ ones} \times 18 = 36 \text{ ones} \\
 & \overline{\;\;\;\;\;\;\;\;\;\;\;5}
\end{array}
$$

$581 \div 18 = 32$ remainder 5

Each of the classrooms receives 32 rulers.
The remaining 5 rulers will be kept in the school office.

2 Ruby worked out the division this way.

```
        3  2  remainder 5
18 ) 5  8  ⁴1
```

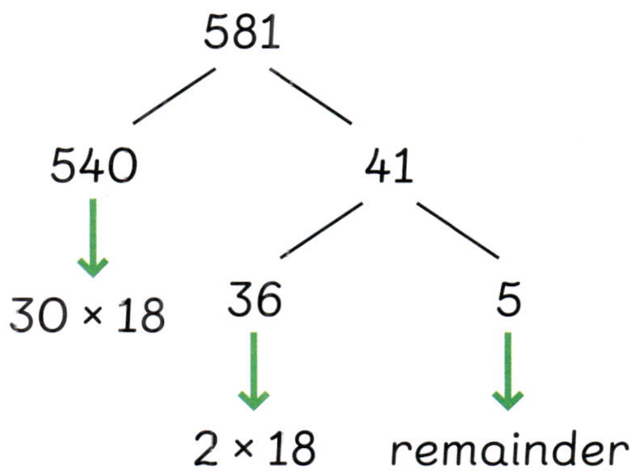

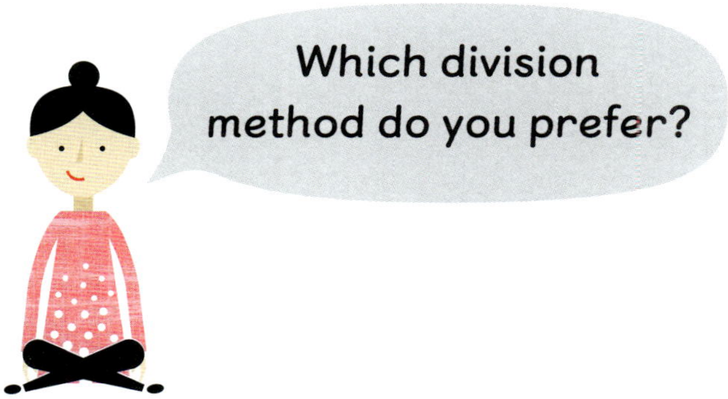

Guided Practice

1 Divide.

(a) 435 ÷ 16 = ☐ remainder ☐

(b) 819 ÷ 17 = ☐ remainder ☐

(c) 546 ÷ 15 = ☐ remainder ☐

2 There are 410 bottles of water on a rack. The bottles are packed in boxes of 14 bottles each. What is the smallest possible number of boxes holding all of the bottles of water?

Complete Worksheet **12**: pages **37–38**

Dividing by a 2-Digit Number with Remainder

Lesson 13

Explore

A farmer used 2004 l of water over 16 days to fill water troughs on her farm. She used an equal volume of water each day.
What is the volume of water she used daily?

Master

1 2004 l ÷ 16 = [ ]

$$\begin{array}{r} 125 \\ 16\overline{)2004} \\ -16 \\ \hline 40 \\ -32 \\ \hline 84 \\ -80 \\ \hline 4 \end{array}$$

2004 ÷ 16 = 125.25
The farmer used 125.25 l of water daily.

2 I can also use short division to find the answer.

$$\begin{array}{r} 1\ 2\ 5\ \text{r}\,4 \\ 16\,\big)\ 2\ 0\ {}^{4}0\ {}^{8}4 \end{array}$$

We can write the remainder 4 as r 4.

$\frac{4}{16} = \frac{1}{4}$

$2004 \div 16 = 125\frac{1}{4}$

The farmer used $125\frac{1}{4}$ l of water daily.

Guided Practice

1 Divide. Show the remainder as a fraction.

(a) $2013 \div 15 =$ ☐ (b) $1916 \div 16 =$ ☐

2 A chain of bakery shops orders flour in bulk. Every month they order 2204 kg of flour. All of the flour is shared equally between 16 shops. What is the mass of flour each shop receives?

3 A group of students collect 3105 kg of old newspapers during a recycling drive. They place them in 18 stacks of equal mass. How heavy is each stack of old newspapers?

Complete Worksheet **13**: pages **39–40**

Solving Word Problems Using Bar Models

Lesson 14

Explore

Amira, Elliott and Ravi collected 234 toys for the school fair.
Elliott collected 16 more toys than Amira.
Ravi collected 3 times as many toys as Elliott.
How many toys did each of them collect?

Master

Understand the problem

Amira collected the fewest toys.
She collected 16 fewer toys than Elliott.
Ravi collected the most toys.

Form a plan

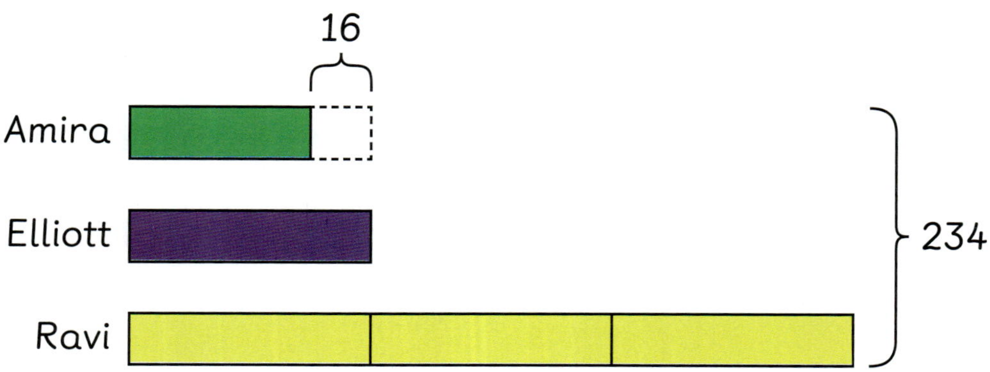

Action the plan

234 + 16 = 250
250 ÷ 5 = 50
1 unit = 50

1 unit − 16 = 50 − 16
= 34
Amira collected 34 toys.

1 unit = 50
Elliott collected 50 toys.

3 units = 3 × 50
= 150
Ravi collected 150 toys.

Guided Practice

1 Lulu used 918 g of flour to make a banana cake, a chocolate cake and a loaf of bread.
She used 4 times as much flour for the bread than the chocolate cake.
She used 30 g less flour for the banana cake than the chocolate cake.

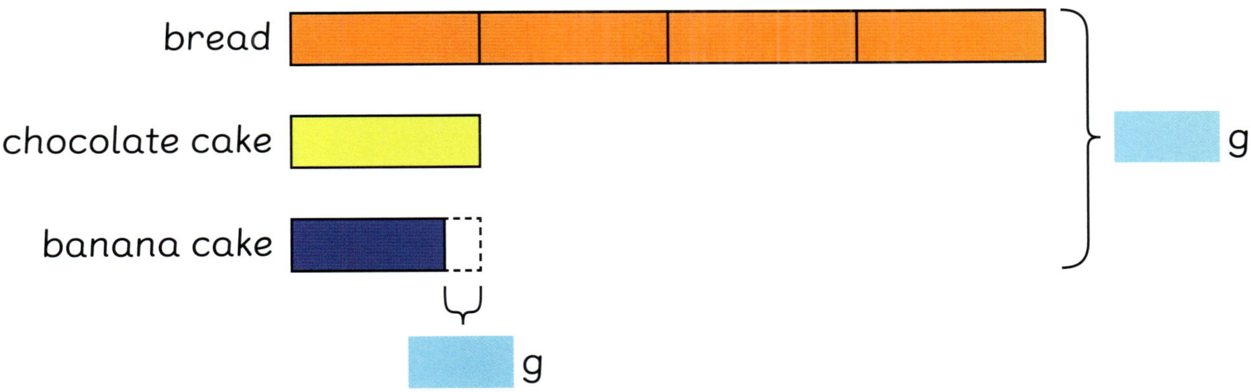

(a) How much more flour did she use for the bread than the chocolate cake?

(b) How much flour did she use for the banana cake?

2 A grocery store sells 23 oranges.
They sell 2 times as many apples as oranges.
They sell 7 more plums than oranges.

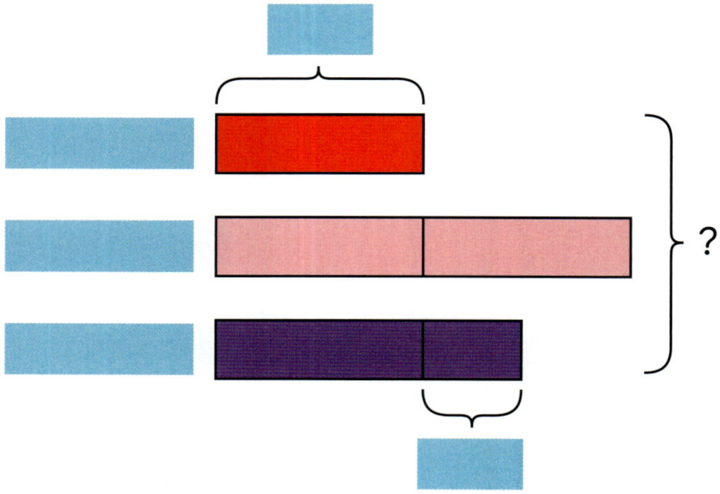

How many pieces of fruit did they sell altogether?

3 A sports centre spends £174 on a cricket bat, a hockey stick and a rugby ball.
The hockey stick cost £10 more than the rugby ball. The cricket bat cost twice as much as the hockey stick.

(a) How much did the hockey stick cost?

(b) How much more did the cricket bat cost than the rugby ball?

Complete Worksheet **14**: pages **41–42**

Solving Word Problems Using Patterns

Lesson 15

Explore

Some pupils are going on a school trip. For every 10 children, there must be 2 adults. All groups of children must have 2 adults supervising them. If there are 449 people altogether on the school trip, what is the smallest possible number of adults there?

Master

Divide to find the number of groups.

$$\begin{array}{r} 3\ 7 \\ 12\overline{)\,4\ 4\ 9} \\ -\ 3\ 6 \\ \hline 8\ 9 \\ -\ \ 8\ 4 \\ \hline 5 \end{array}$$

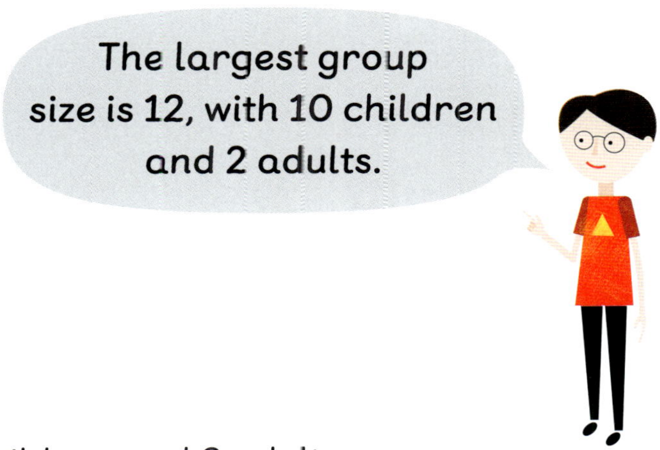

There are 37 groups of 10 children and 2 adults.
There must be 2 adults in the final group of 5 children.

$37 \times 2 =$ ____

____ $+ 2 =$ ____

The smallest number of adults on the trip is ____ .

Guided Practice

1 Teachers at Sam's school are setting up the hall for a quiz afternoon. They need 4 highlighters and 9 blue pens on each table. If there are 360 pens altogether, what is the greatest number of tables that can have the correct number of pens?

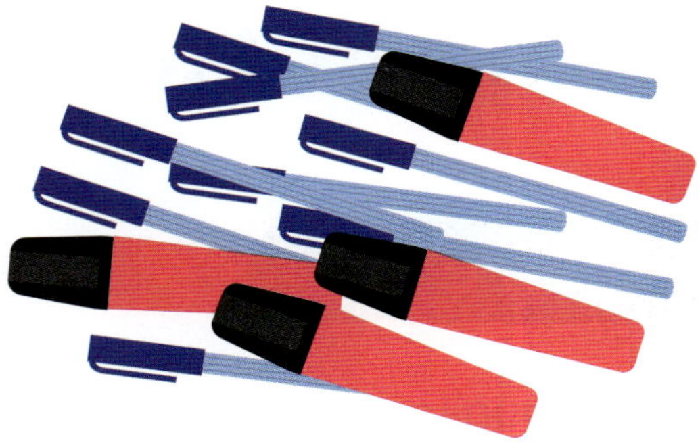

2 A book needs an illustration on the front cover, the back cover and on every 10th page in the book.

(a) If the book had 200 pages, how many illustrations would there be in total?

(b) If the book had 27 illustrations, what is the smallest possible number of pages in the book?

Complete Worksheet **15**: pages **43–44**

Solving Word Problems Using Multiple Methods

Lesson 16

Explore

A company produces stickers and sells them in packets.
Each packet of stickers contains 14 regular stickers and 12 shiny stickers.
How many packets of stickers can be made from 1921 regular stickers and 1670 shiny stickers?

Master

1 1921 ÷ 14 = ☐

$$\begin{array}{r} 137 \text{ remainder } 3 \\ 14\overline{)1921} \\ -14 \\ \hline 52 \\ -42 \\ \hline 101 \\ 98 \\ \hline 3 \end{array}$$

2 $1670 \div 12 = $ ▢

$$\begin{array}{r} 1\ 3\ 9 \text{ remainder } 2 \\ 12 \overline{) \ 1\ 6\ {}^{4}7\ {}^{11}0} \end{array}$$

There are 139 groups of 12 shiny stickers with 2 left over.

1670

1200, 360, 108, 2

1200 ÷ 12 → 100

360 ÷ 12 → 30

108 ÷ 12 → 9

This is the remainder.

The greatest number of packets that can be made is 137. There are not enough regular stickers to make 139 packets.

3. How many more of each type of sticker are needed to make exactly 140 packets with no stickers left over?

$$140 \times 14 = (140 \times 10) + (140 \times 4)$$
$$= 1400 + 560$$
$$= 1960$$

1960 regular stickers are needed for 140 packets.

$$140 \times 12 = (140 \times 10) + (140 \times 2)$$
$$= 1400 + 280$$
$$= 1680$$

1680 shiny stickers are needed for 140 packets.

$1960 - 1921 = 39$
$1680 - 1670 = 10$

39 more regular stickers and 10 more shiny stickers are needed to make 140 packets with no stickers left over.

Guided Practice

Solve.
13 screws and 16 nails are used to make a birdhouse.
A factory has 2057 screws and 2513 nails.

1. What is the greatest number of birdhouses that the factory can make?

2. How many nails and screws are needed to make 170 birdhouses?

Complete Worksheet **16**: pages **45–46**

Finding Common Multiples

Lesson 17

Explore

A shop sells blue pens in packets of 12 and black pens in packets of 14.
Lulu wants to buy an equal number of blue pens and black pens.
What is the smallest number of packets of each colour of pen should she buy?

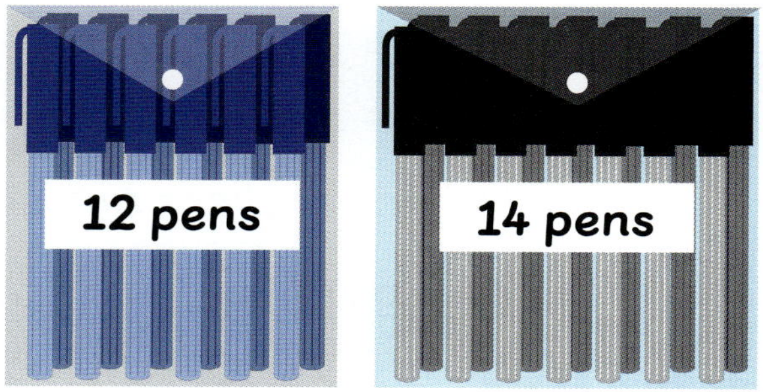

Master

1

Number of packets	Total number of blue pens
1	12
2	24
3	36
4	48
5	60
6	72
7	84

Number of packets	Total number of black pens
1	14
2	28
3	42
3	56
5	70
6	84
7	96

Lulu should buy 7 packets of blue pens and 6 packets of black pens to have an equal number of each colour of pen.

2 $7 \times 12 = 84$ $6 \times 14 = 84$

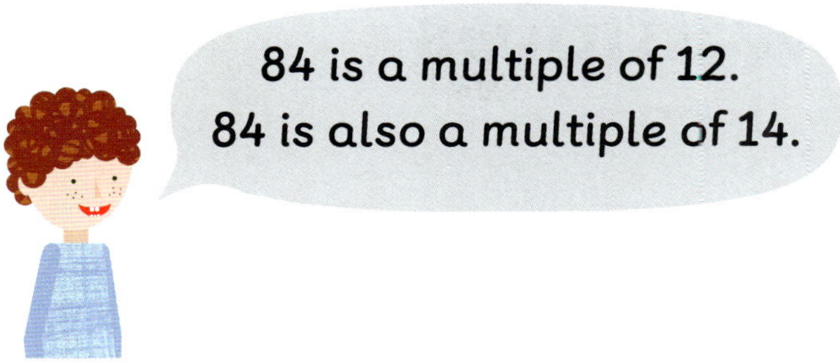

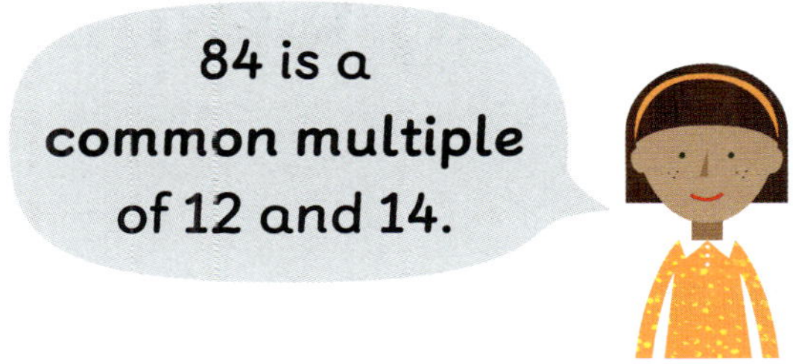

3 Find the first 3 common multiples of 5 and 10.

Multiples of 5: 5, 10, 15, 20, 25, 30, 35, 40, ...
Multiples of 10: 10, 20, 30, 40, 50, 60, 70, 80, ...

The first 3 common multiples of 5 and 10 are 10, 20 and 30.

Guided Practice

1 Find the first three common multiples of:

(a) 3 and 6

(b) 4 and 6

(c) 5 and 6

(d) 8 and 6

(e) 12 and 15

2 A group of children play dodgeball and rounders at an indoor sports arena. They are first put into teams of 12 for dodgeball and then into teams of 16 for rounders. What is the minimum number of children at the sports arena if they can be split into teams with no children left over?

Complete Worksheet **17**: pages **47–48**

Finding Common Multiples

Lesson 18

Explore

At a school sports event, children are grouped for each sport.

tennis:
groups of 4

volleyball:
groups of 6

tug of war:
groups of 8

Once this is done, every child is in a group and all groups have the correct number of children.

How many children could there be at the sports event?

Master

1

Multiples of 4: 4, 8, 12, 16, 20, 24, 28, 32, 36, 40

2

Multiples of 6: 6, 12, 18, 24, 30, 36, 42, 48, 54, 60

3

Multiples of 8: 8, 16, 24, 32, 40, 48, 56, 64, 72, 80

4 Use a table to show the multiples.

Multiples of 4	4	8	12	16	20	24	28	32	36	40	44	48
Multiples of 6	6	12	18	24	30	36	42	48	54	60	66	72
Multiples of 8	8	16	24	32	40	48	56	64	72	80	88	96

24 and 48 are common multiples of 4, 6 and 8. There could be 24 or 48 children at the sports event.

Guided Practice

1 Complete the table. Circle two common multiples of 5 and 7.

Multiples of 5														
Multiples of 7														

2 Find two common multiples of:

(a) 3 and 5

(b) 9 and 4

(c) 3, 6 and 5

(d) 4, 5 and 8

3 A baker makes a batch of bread rolls. The bread rolls can be put exactly into bags of 4, 8 or 9. What is the smallest number of bread rolls the baker could have made?

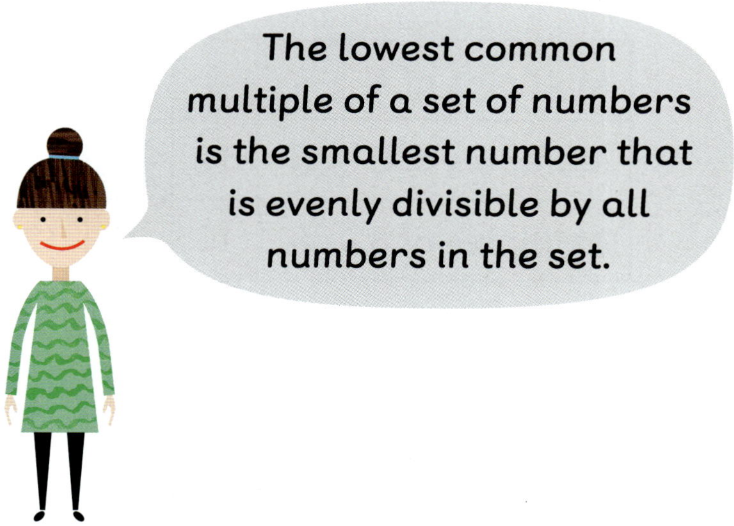

Complete Worksheet **18**: pages **49–50**

Finding Common Factors

Lesson 19

Explore

A theme park is designing a new ride. It wants the ride to hold exactly 168 or 192 passengers at one time. Each carriage should hold at least 10 passengers, but no more than 15 passengers.
How many seats should each carriage have so that the ride can hold exactly 168 or 192 passengers?

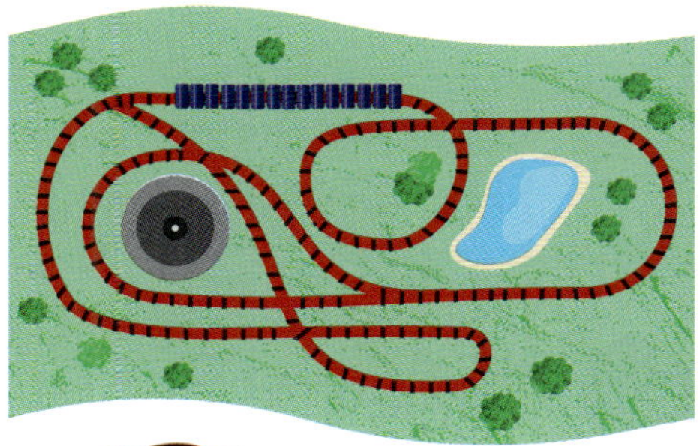

What are the factors of 168 and 192?

Master

I know 10 is not a factor of 168 and 192.

I know 15 is not a factor of 168 and 192.

Carriages of 10 or 15 cannot hold 168 or 192 passengers exactly.

1 Is 14 a factor of 168?

```
         1  2
     ---------
14 )  1  6  8
    - 1  4      →  14 tens ÷ 14 = 1 ten
    ---------
         2  8
    -    2  8   →  18 ones ÷ 14 = 2 ones
    ---------
            0
```

14 is a factor of 168.

Is 14 a factor of 192?

$$\begin{array}{r|lll} & & \square & \square \\ \hline 14 & 1 & 9 & 2 \\ - & 1 & 4 & \\ \hline & & 5 & 2 \\ - & & 4 & 2 \\ \hline & & 1 & 0 \end{array}$$

→ 14 tens ÷ 14 = ☐ ten

→ 42 ones ÷ 14 = ☐ ones

Carriages of 14 can hold 168 passengers but not 192 passengers exactly.

2 Is 13 a factor of 168?

$$\begin{array}{r|lll} & & \square & \square \\ \hline 13 & 1 & 6 & 8 \\ - & 1 & 3 & \\ \hline & & 3 & 8 \\ - & & 2 & 6 \\ \hline & & 1 & 2 \end{array}$$

168

130 38

13 is not a factor of 168.

3 Is 12 a factor of 168?

$$\begin{array}{r|lll} & & \square & \square \\ \hline 12 & 1 & 6 & 8 \\ - & 1 & 2 & \\ \hline & & 4 & 8 \\ - & & 4 & 8 \\ \hline & & & 0 \end{array}$$

168

120 48

12 is a factor of 168.

Is 12 a factor of 192?

```
        ▢ ▢
    _______
12 ) 1 9 2
   - 1 2
   -------
       7 2
   -   7 2
   -------
         0
```

```
      192
     /   \
  120     72
```

12 is a factor of 192.
12 is a common factor of 168 and 192.
Each carriage should have 12 seats to hold 168 or 192 passengers exactly.

4 Find the common factors of 72 and 88.

Factors of 72: 1, 2, 3, 4, 6, 8, 9, 12, 18, 24, 36, 72

Factors of 88: 1, 2, 4, 8, 11, 22, 44, 88

1, 2, 4 and 8 are common factors of 72 and 88.

8 is the greatest common factor of 72 and 88.

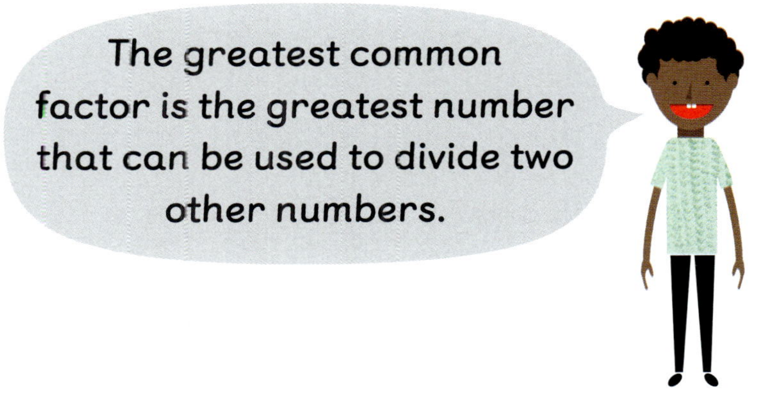

Guided Practice

1 Find all the common factors of 136 and 144.

Factors of 136: ___, ___, ___, ___, ___, ___, ___, ___, ___, ___

Factors of 144: ___, ___, ___, ___, ___, ___, ___, ___, ___, ___, ___, ___, ___, ___, ___

The common factors of both 136 and 144 are ___, ___, ___, ___, ___ and ___.

2 Find the common factors of:

(a) 32 and 48

(b) 34 and 51

3 Find the greatest common factor of:

(a) 98 and 51

(b) 105 and 133

Complete Worksheet **19**: pages **51–52**

Finding Common Factors

Lesson 20

Explore

A garden centre is deciding on how to arrange and stack bags of compost.
Each layer needs 18 bags of compost and needs to be arranged in the shape of a rectangle.
How can this be done?

Master

1

1 row of 18 bags
1 × 18 = 18

2 rows of 9 bags
2 × 9 = 18

3 rows of 6 bags
3 × 6 = 18

1, 2, 3, 6, 9 and 18 are factors of 18.

2 Find the common factors of 8, 16 and 24.

Factors of 8: 1, 2, 4, 8
Factors of 16: 1, 2, 4, 8, 16
Factors of 24: 1, 2, 3, 4, 6, 8, 12, 24

All whole numbers have 1 as a factor.

8, 16 and 24 have the common factors of 1, 2, 4 and 8.

All even numbers have 2 as a factor.

3 Use a table to show the common factors of 90, 105 and 120.

	Factors															
90	1	2	3	5	6	9	10	15	18	30	45	90				
105	1	3	5	7	15	21	35	105								
120	1	2	3	4	5	6	8	10	12	15	20	24	30	40	60	120

1, 3, 5 and 15 are common factors of 90, 105 and 120.
15 is the greatest common factor of 90, 105 and 120.

Guided Practice

1 Complete the table. Find the common factors.

Factors of 28										
Factors of 42										
Factors of 80										

2 What are the two greatest 2-digit numbers that have the common factors 6, 7 and 8?

Complete Worksheet **20**: pages **53–54**

Finding Prime Numbers

Lesson 21

Explore

Elliott has 7 square tiles.

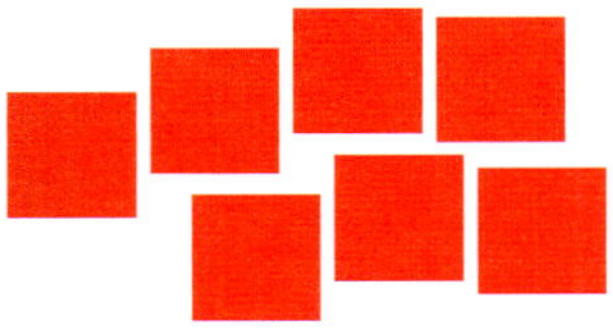

Charles has 12 square tiles.

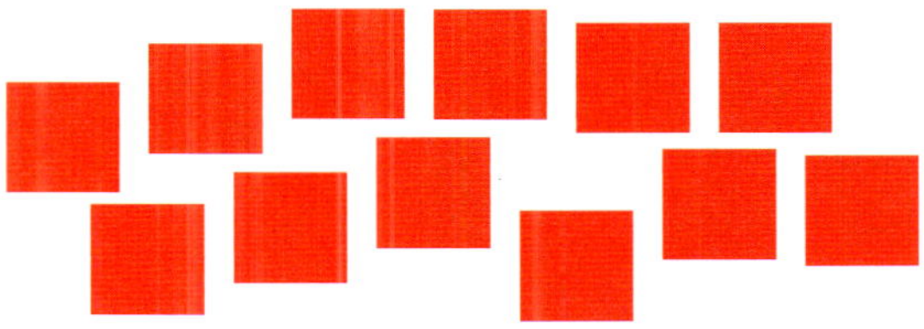

How many different rectangular arrangements can each of them make with their tiles?

Master

1 Elliott can only make 1 rectangular arrangement.

1 row of 7

1 × 7 = 7

The factors of 7 are 1 and 7.

7 is a **prime number**.

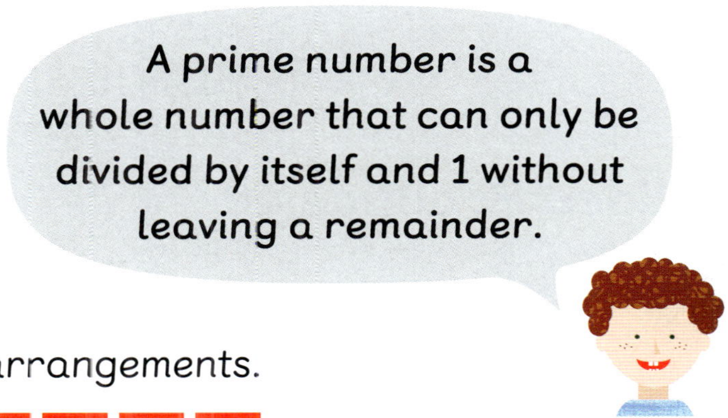

2 Charles can make 3 rectangular arrangements.

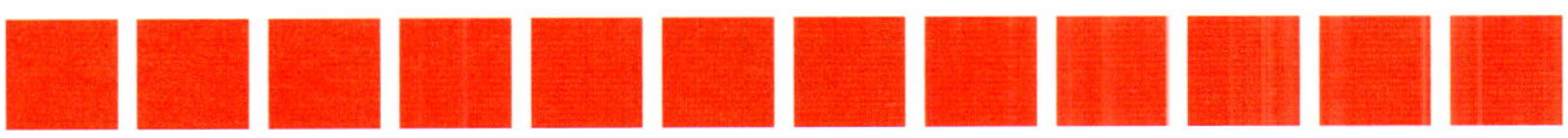

1 × 12 = 12

2 × 6 = 12

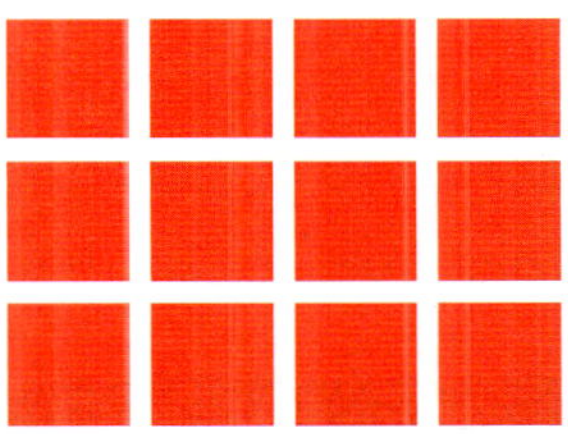

3 × 4 = 12

The factors of 12 are 1, 2, 3, 4, 6 and 12.
12 is a **composite number**.

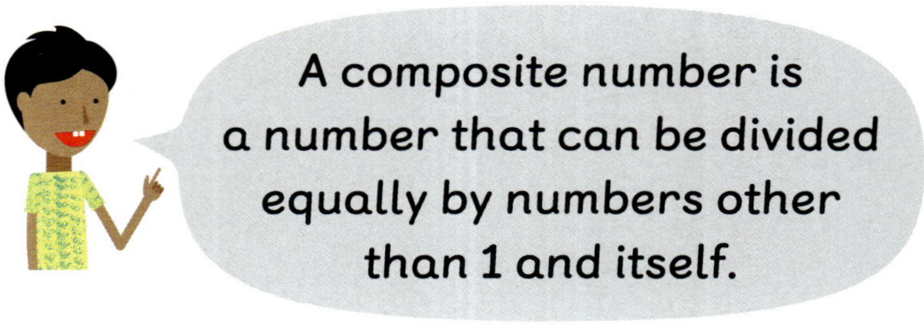

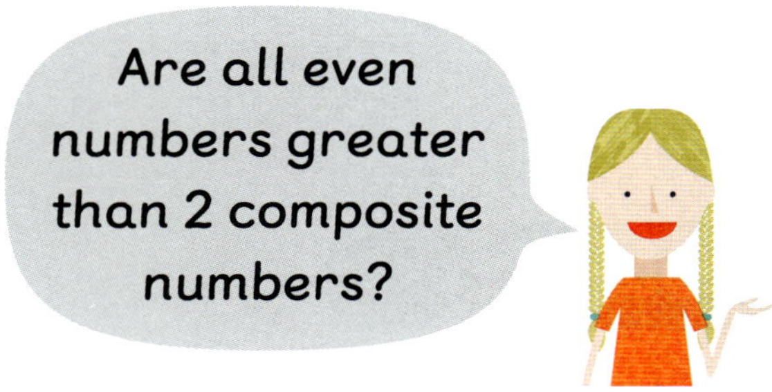

3 Are the following numbers prime or composite?

13	15
$1 \times 13 = 13$	$1 \times 15 = 15$
$13 \times 1 = 13$	$3 \times 5 = 15$
	$5 \times 3 = 15$
	$15 \times 1 = 15$
$13 \div 1 = 13$	$15 \div 1 = 15$
$13 \div 13 = 1$	$15 \div 3 = 5$
	$15 \div 5 = 3$
	$15 \div 15 = 1$

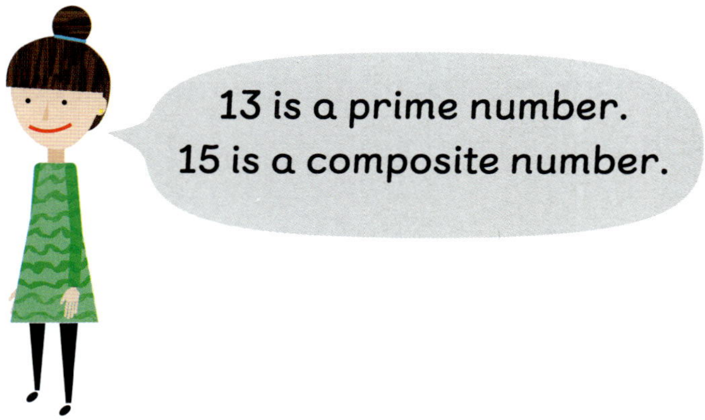

Guided Practice

1 Look at the following numbers. Which are prime? Which are composite?

55	57	64	96	97	113	128	129	147	151

2 List all the prime numbers between 65 and 95.

3 Elliott thinks that 1048 is a composite number.
How can Elliott check whether he is correct?

Complete Worksheet **21**: page **55**

Finding Prime Numbers

Lesson 22

Explore

Oak makes this rectangular arrangement using coloured tiles.
What do you notice about the number of tiles of each colour she uses?

Master

1

8 = 5 + 3

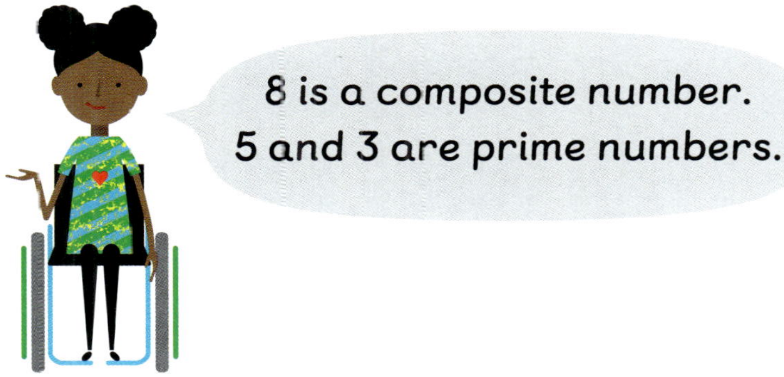

2

10 = 7 + 3

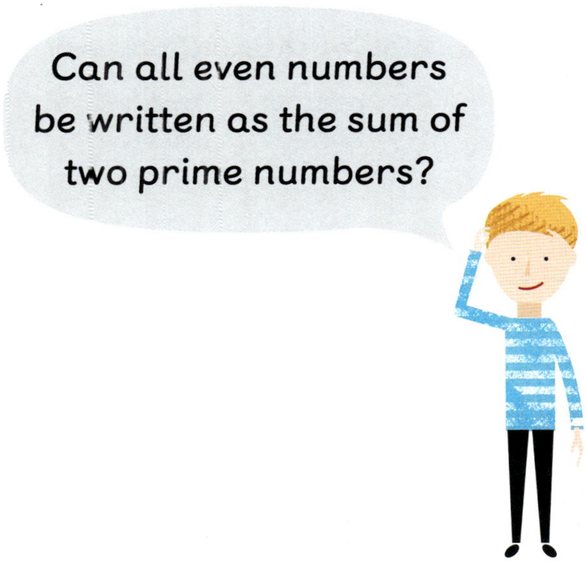

3

16 = 11 + 5

4 Show each even number as the sum of two prime numbers.

18 = 11 + 7

20 = 13 + 7

22 = 19 + 3

24 = ☐ + ☐

26 = ☐ + ☐

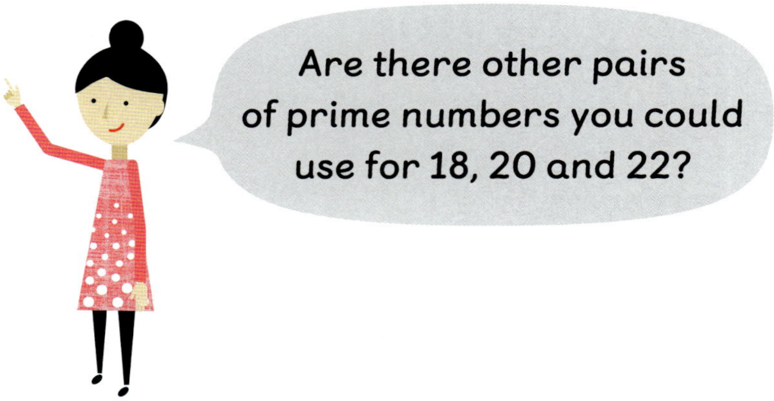

Guided Practice

1 Show each even number as the sum of two prime numbers.

(a) 12 = ☐ + ☐

(b) 14 = ☐ + ☐

(c) 28 = ☐ + ☐

(d) 40 = ☐ + ☐

2 Write 50 as the sum of two prime numbers that have the smallest difference.

50 = ☐ + ☐

Complete Worksheet **22**: page **56**

Mind Challenge

We can find all the prime numbers up to 100 using the sieve of Eratosthenes. On a 100-square, shade all the multiples of 2, 3, 5, 7 and 11, but don't shade the numbers 2, 3, 5, 7 and 11 themselves.
What numbers remain unshaded?
Are these all prime numbers?

1	2	3	4	5	6	7	8	9	10
11	12	13	14	15	16	17	18	19	20
21	22	23	24	25	26	27	28	29	30
31	32	33	34	35	36	37	38	39	40
41	42	43	44	45	46	47	48	49	50
51	52	53	54	55	56	57	58	59	60
61	62	63	64	65	66	67	68	69	70
71	72	73	74	75	76	77	78	79	80
81	82	83	84	85	86	87	88	89	90
91	92	93	94	95	96	97	98	99	100

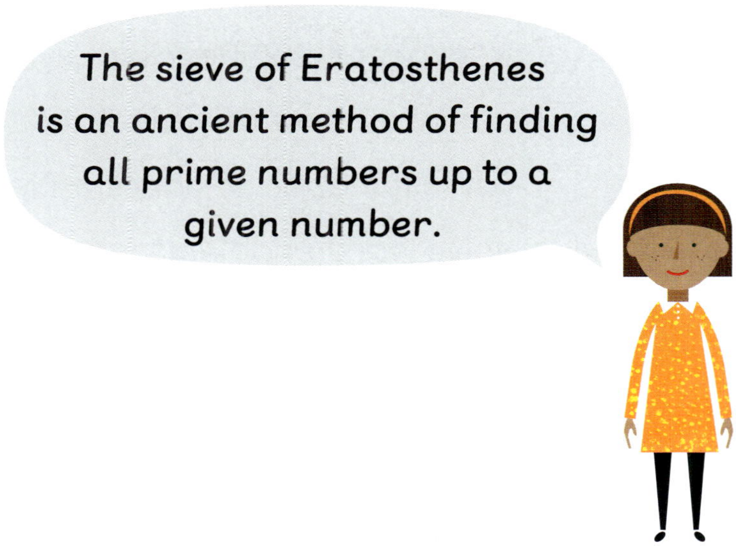

My Maths Journal

Without using a calculator, find the quickest way to:

- add 19 to a number.
- subtract 99 from a number.
- multiply a number by 8.
- divide a number by 25.
- multiply a number by 99.

Review

I can...

- ☐ recognise when and how to use the order of operations.
- ☐ multiply numbers by tens.
- ☐ multiply numbers up to 4 digits by a 2-digit whole number.
- ☐ divide numbers up to 4 digits by a 2-digit whole number.
- ☐ interpret remainders in division calculations.
- ☐ carry out mental calculations.
- ☐ use estimation to check answers.
- ☐ solve word problems involving the four operations using bar models, patterns and other methods.
- ☐ find multiples of a number and identify common multiples of a set of numbers.
- ☐ find the lowest common multiple of a set of numbers.
- ☐ find factors of a number and identify common factors of a set of numbers.
- ☐ find the greatest common factor of a set of numbers.
- ☐ solve word problems using multiples and factors.
- ☐ identify prime numbers and composite numbers.

Can you find $\frac{1}{4}$?

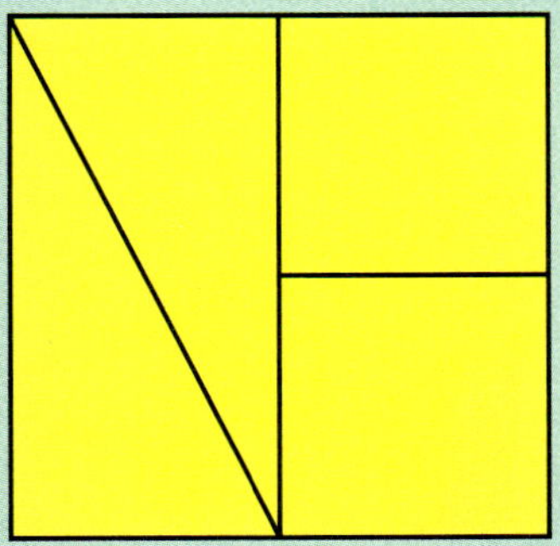

Can you find $\frac{1}{3}$?

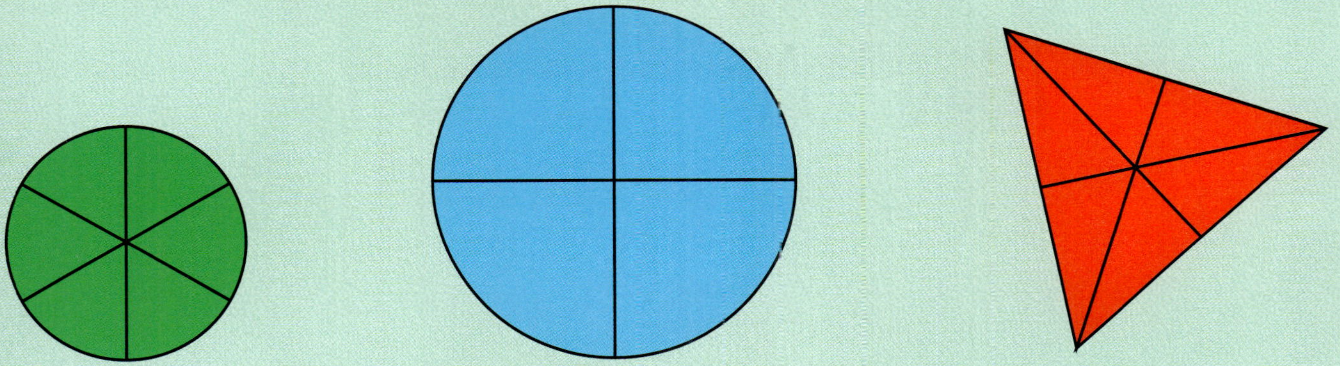

What other fractions can you see?

Chapter 3
Fractions

Simplifying Fractions Using Common Factors

Lesson 1

Explore

I cut out $\frac{1}{4}$ of my strip of paper.

I cut out $\frac{5}{20}$ of my strip of paper.

I cut out $\frac{3}{12}$ of my strip of paper.

I cut out $\frac{2}{8}$ of my strip of paper.

Do they all cut out the same length from their strips of paper?
How can you tell?

Master

1  folded his strip into 4 equal parts and cut out 1 part.

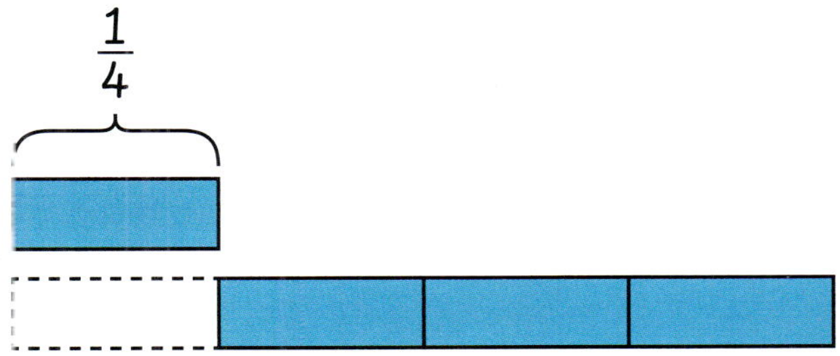

 folded her strip into 12 equal parts and cut out 3 parts.

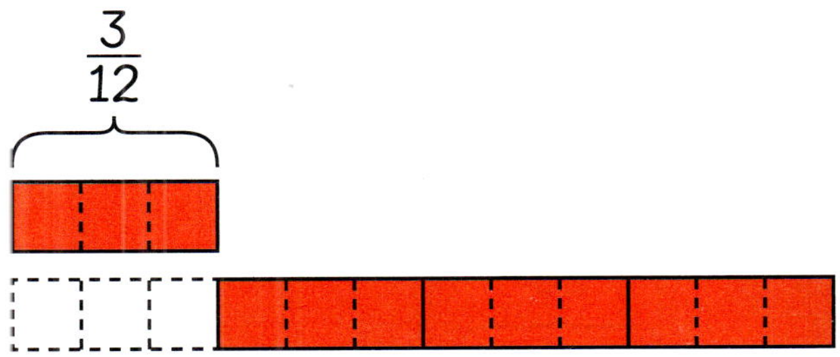

 folded her strip into 20 equal parts and cut out 5 parts.

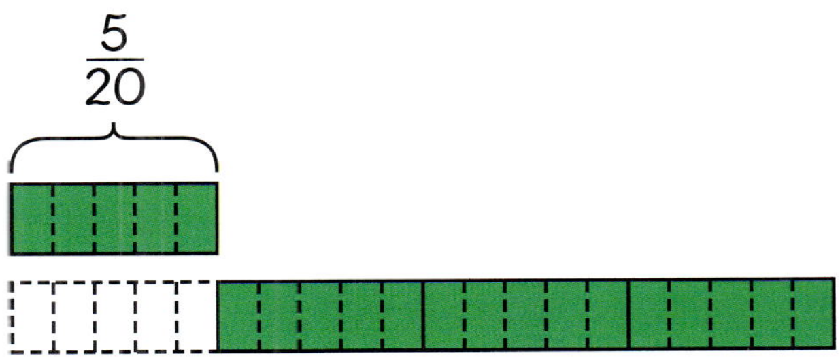

 folded his strip into 8 equal parts and cut out 2 parts.

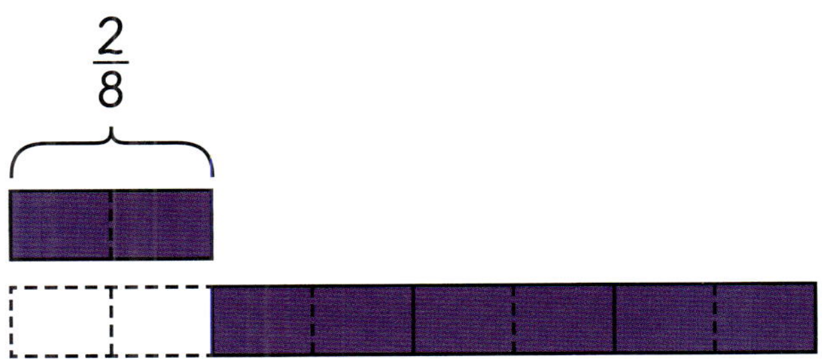

They all cut out the same length from their strips of paper.

2

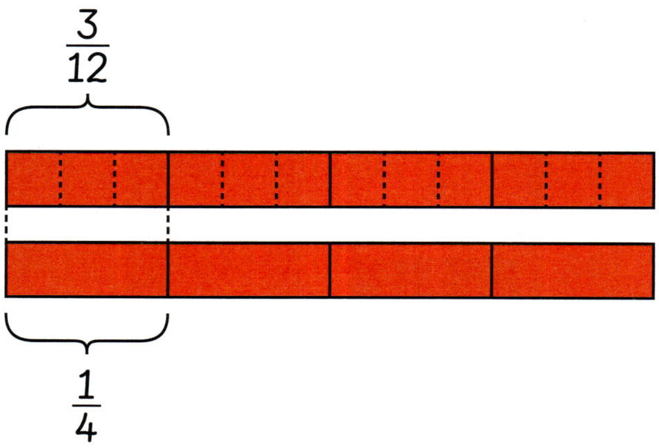

$3 = 1 \times 3 \qquad 12 = 1 \times 12$

$12 = 2 \times 6$

$12 = 3 \times 4$

$$\frac{3}{12} = \frac{1}{4} \quad (\div 3 \text{ numerator and denominator})$$

3 is a common factor of 3 and 12.

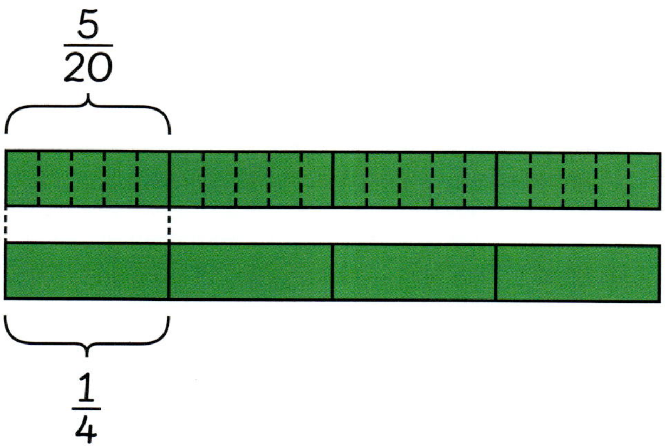

$5 = 1 \times 5 \qquad 20 = 1 \times 20$

$= 2 \times 10$

$= 4 \times 5$

$$\frac{5}{20} = \frac{1}{4} \quad (\div 5 \text{ numerator and denominator})$$

5 is a common factor of 5 and 20.

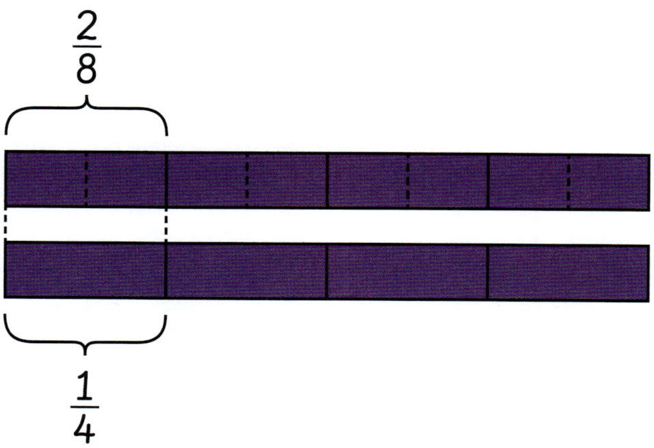

$2 = 1 \times 2 \qquad 8 = 1 \times 8$

$8 = 2 \times 4$

$$\frac{2}{8} = \frac{1}{4} \quad (\div 2 \text{ numerator and denominator})$$

2 is a common factor of 2 and 8.

Guided Practice

1 Simplify.

(a) 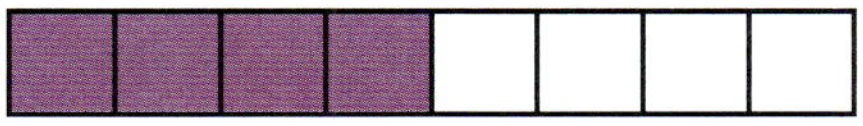

$\frac{4}{8} = \frac{\square}{\square}$

$\div \square$

$\frac{4}{8} = \frac{\square}{\square}$

$\div \square$

(b) 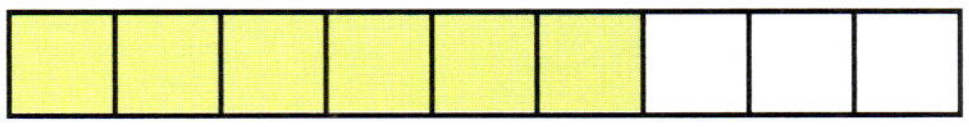

$\frac{6}{9} = \frac{\square}{\square}$

$\div \square$

$\frac{6}{9} = \frac{\square}{\square}$

$\div \square$

(c) 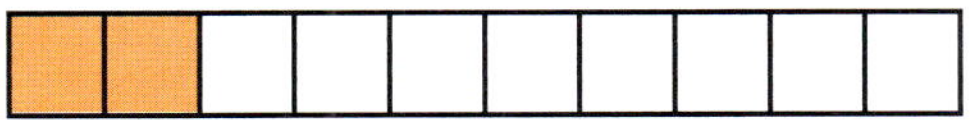

$\frac{2}{10} = \frac{\square}{\square}$

$\div \square$

$\frac{2}{10} = \frac{\square}{\square}$

$\div \square$

2 Write each fraction in its simplest form.

(a) $\frac{2}{4} = \frac{\square}{\square}$ (b) $\frac{3}{6} = \frac{\square}{\square}$ (c) $\frac{5}{10} = \frac{\square}{\square}$

(d) $\frac{2}{6} = \frac{\square}{\square}$ (e) $\frac{10}{12} = \frac{\square}{\square}$ (f) $\frac{3}{9} = \frac{\square}{\square}$

Complete Worksheet **1**: pages **61–62**

Simplifying Fractions Using Common Factors

Lesson 2

Explore

Master

1 $\frac{8}{12} = \frac{\square}{6}$

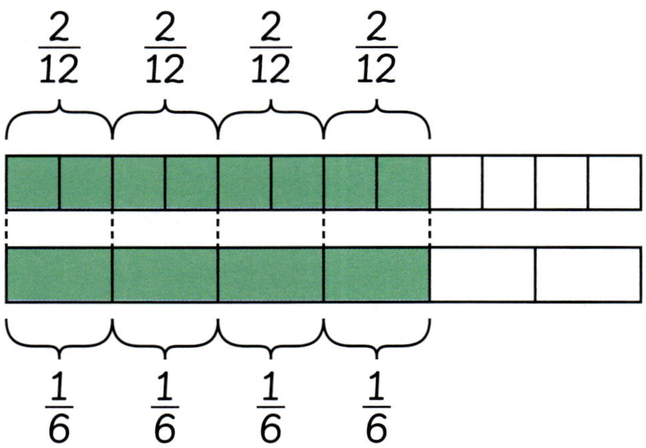

The egg box is $\frac{4}{6}$ full.

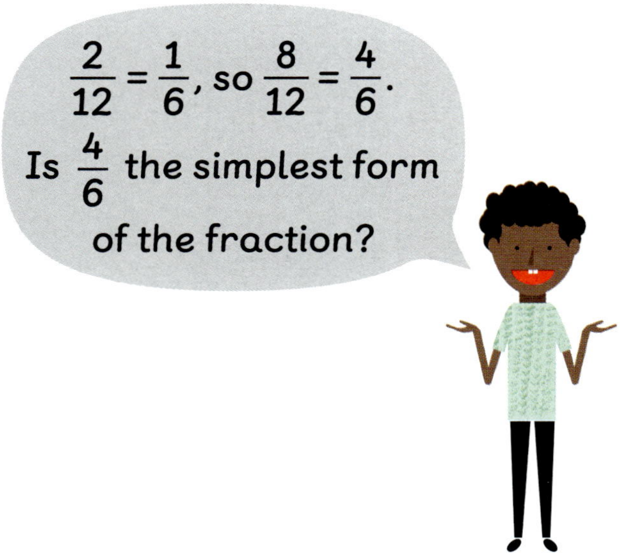

2 $\frac{8}{12} = \frac{\square}{3}$

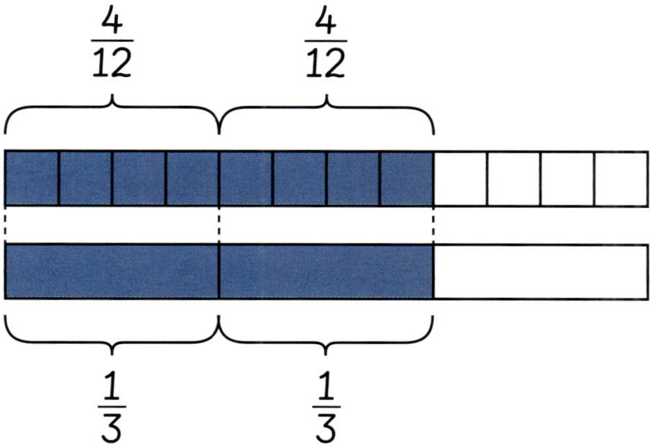

The egg box is $\frac{2}{3}$ *full.*

3 Amira used 2 of the eggs. What fraction of the egg box is filled now?

$\frac{6}{12} = \frac{\square}{6} = \frac{\square}{2}$

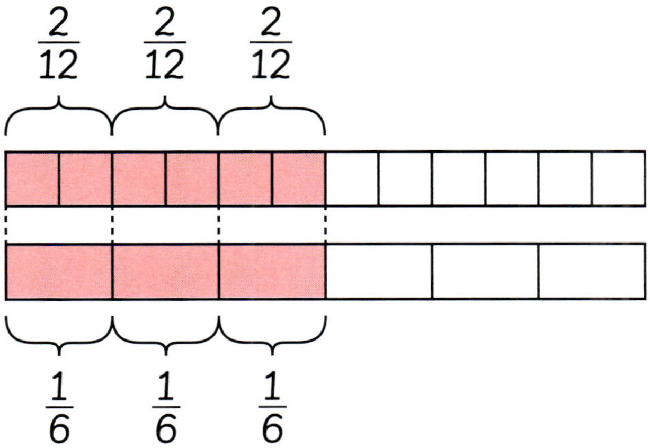

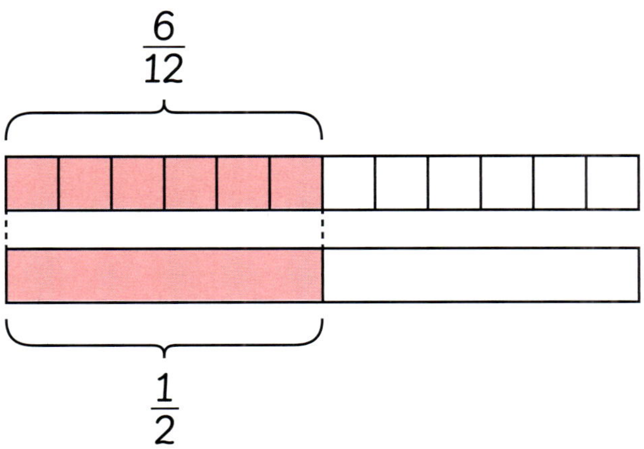

The egg box is $\frac{1}{2}$ *full now.*

Guided Practice

1 Write the simplest form of each fraction.

(a)

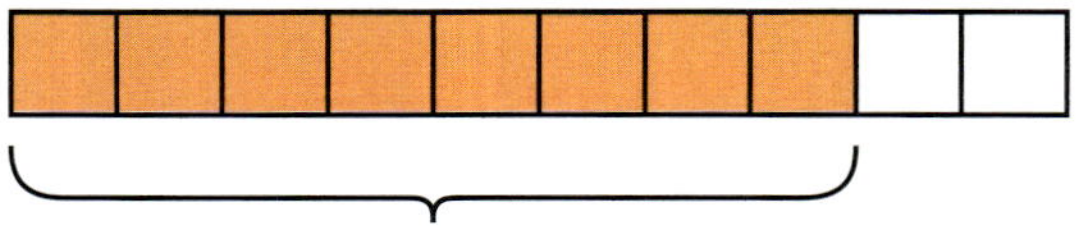

$$\frac{8}{10} = \frac{\square}{\square}$$

(b)

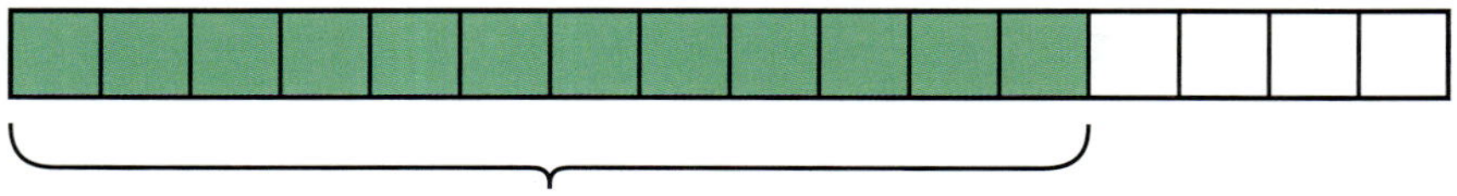

$$\frac{12}{16} = \frac{\square}{\square}$$

2 Which of the fractions below are in the simplest form?

$\frac{1}{12}$	$\frac{5}{12}$	$\frac{9}{12}$
$\frac{2}{12}$	$\frac{6}{12}$	$\frac{10}{12}$
$\frac{3}{12}$	$\frac{7}{12}$	$\frac{11}{12}$
$\frac{4}{12}$	$\frac{8}{12}$	$\frac{12}{12}$

Complete Worksheet **2**: pages **63–64**

Comparing and Ordering Proper Fractions

Lesson 3

Explore

Lulu, Holly and Charles share a bar of chocolate.

Lulu takes $\frac{1}{3}$ of the bar, Holly takes $\frac{5}{12}$ and Charles takes $\frac{1}{4}$.

Who takes the most chocolate?

Master

1 Compare using bar models.

 $\frac{1}{3}$ 

 $\frac{5}{12}$ 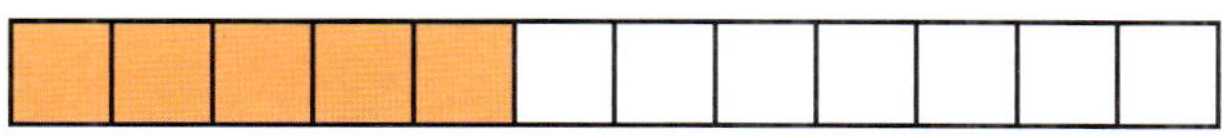

 $\frac{1}{4}$ 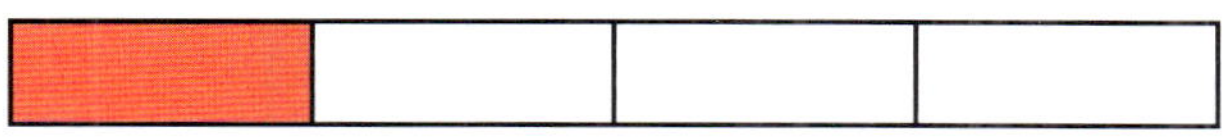

Holly takes the most chocolate.

2 Put $\frac{1}{3}$, $\frac{5}{12}$ and $\frac{1}{4}$ in order from the smallest to the greatest.

$$\frac{1}{3} \overset{\times 4}{=} \frac{4}{12}$$

$$\frac{1}{4} \overset{\times 3}{=} \frac{3}{12}$$

$$\frac{5}{12}$$

$$\frac{3}{12} < \frac{4}{12} < \frac{5}{12}$$

$$\frac{1}{4} < \frac{1}{3} < \frac{5}{12}$$

We can easily compare fractions with the same denominator.

$\frac{1}{4}$ $\frac{1}{3}$ $\frac{5}{12}$

smallest → greatest

Holly takes the most chocolate. Charles takes the least chocolate.

Guided Practice

1 Which of these is the greatest fraction?

$\frac{5}{12}$ $\frac{1}{2}$ $\frac{3}{4}$

2 Which of these is the smallest fraction?

$\frac{2}{3}$ $\frac{5}{6}$ $\frac{15}{18}$

3 Order these fractions from the greatest to the smallest.

$\frac{2}{3}$ $\frac{11}{15}$ $\frac{3}{5}$

Complete Worksheet **3**: pages **65–66**

Comparing and Ordering Improper Fractions

Lesson 4

Explore

Ruby, Oak and Sam are sharing some equal-sized tarts.

Who takes the most? Who takes the least?

Master

1 Compare using bar models.

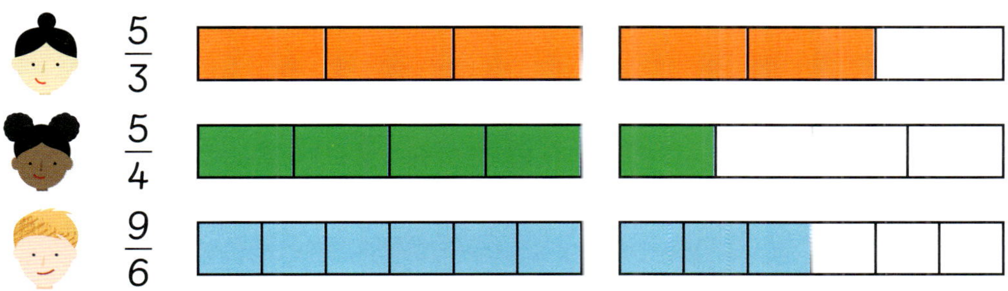

Ruby takes the most. Oak takes the least.

2 Put $\frac{5}{3}$, $\frac{5}{4}$ and $\frac{9}{6}$ in order from the greatest to the smallest.

$$\frac{5}{3} \overset{\times 4}{=} \frac{20}{12}$$

$$\frac{5}{4} \overset{\times 3}{=} \frac{15}{12}$$

$$\frac{9}{6} \overset{\times 2}{=} \frac{18}{12}$$

$$\frac{20}{12} > \frac{18}{12} > \frac{15}{12}$$

$$\frac{5}{3} > \frac{9}{6} > \frac{5}{4}$$

Compare by making the denominator the same. 12 is the lowest common multiple of 3, 4 and 6.

$\frac{5}{3}$ $\frac{9}{6}$ $\frac{5}{4}$

greatest → smallest

Guided Practice

1 Which of these is the greatest fraction?

$\frac{5}{2}$ $\frac{13}{4}$ $\frac{11}{6}$

2 Which of these is the smallest fraction?

$\frac{7}{5}$ $\frac{7}{2}$ $\frac{9}{4}$

3 Order these fractions from the smallest to the greatest.

$\frac{17}{9}$ $\frac{11}{6}$ $\frac{3}{2}$

Find the lowest common multiple of the 3 different denominators. Multiply so that all 3 fractions have the same denominator.

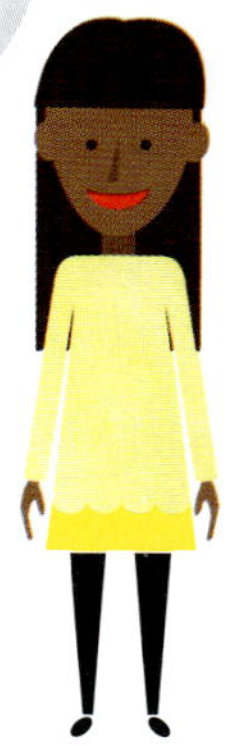

Complete Worksheet **4**: pages **67–68**

Comparing and Ordering Fractions and Mixed Numbers

Lesson 5

Explore

Sam is comparing recipes for a green curry.
Which recipe uses the greatest amount of coconut milk?

Cathy's Green Curry

- 1 tbsp vegetable oil
- 1 tbsp garlic and ginger paste
- 6 tbsp green curry paste
- 2 tbsp fish sauce
- $\frac{11}{15}$ cup coconut milk

Anna's Green Curry

- 1 tbsp vegetable oil
- 1 tbsp garlic and ginger paste
- 6 tbsp green curry paste
- 2 tbsp fish sauce
- $1\frac{2}{3}$ cups coconut milk

Varsha's Green Curry

- 1 tbsp vegetable oil
- 1 tbsp garlic and ginger paste
- 6 tbsp green curry paste
- 2 tbsp fish sauce
- $1\frac{2}{5}$ cups coconut milk

Master

1 Compare $\frac{11}{15}$ with $1\frac{2}{3}$ and $1\frac{2}{5}$.

$\frac{11}{15} < 1\frac{2}{3}$ $\frac{11}{15} < 1\frac{2}{5}$

$\frac{11}{15}$ cup is the smallest amount of coconut milk.

$\frac{11}{15}$ is less than 1.

2 Compare $1\frac{2}{3}$ and $1\frac{2}{5}$.

$1\frac{2}{3} = 1\frac{10}{15}$ $1\frac{2}{5} = 1\frac{6}{15}$

$1\frac{10}{15} > 1\frac{6}{15}$

$1\frac{2}{3} > 1\frac{2}{5}$

$1\frac{2}{3}$ cup is the greatest amount of coconut milk.

Find a common denominator. What is the lowest common multiple of 3 and 5?

3 Compare $\frac{11}{15}$, $1\frac{2}{3}$ and $1\frac{2}{5}$.

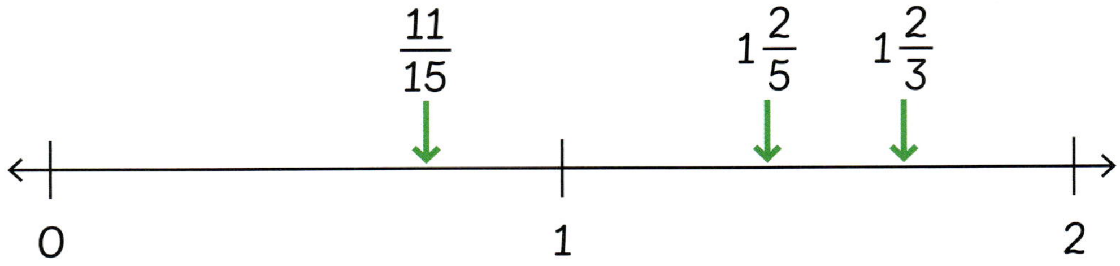

$\frac{11}{15} < 1\frac{2}{5} < 1\frac{2}{3}$

Guided Practice

1 Find a common denominator and compare using > or <.

(a) $1\frac{1}{2}$ ☐ $1\frac{1}{5}$ (b) $2\frac{2}{3}$ ☐ $2\frac{3}{4}$

2 Put the following numbers in order from the smallest to the greatest.

(a) $\frac{7}{3}$ $4\frac{1}{6}$ $3\frac{2}{5}$

(b) $3\frac{3}{4}$ $2\frac{3}{8}$ $\frac{15}{6}$

3 Place $1\frac{3}{5}$, $1\frac{17}{20}$ and $\frac{13}{8}$ on the number line.

Arrange them in order from the smallest to the greatest.

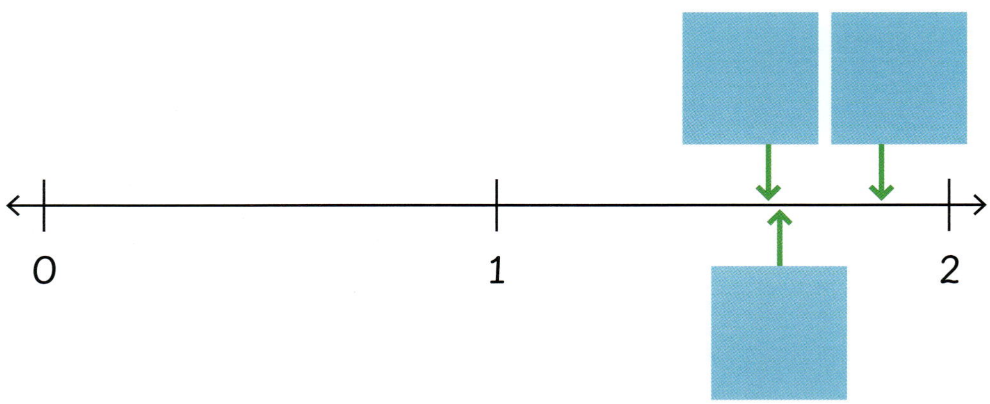

Complete Worksheet **5**: pages **69–70**

Adding and Subtracting Unlike Fractions

Lesson 6

Explore

Amira has $\frac{1}{2}$ l of juice and Elliott has $\frac{1}{3}$ l of juice.

Is Charles correct?

Master

1 Is $\frac{1}{2} + \frac{1}{3} = \frac{2}{5}$?

Method 1

$\frac{1}{2} > \frac{1}{5}$ and $\frac{1}{3} > \frac{1}{5}$.

So, $\frac{1}{2} + \frac{1}{3}$ must be greater than $\frac{2}{5}$.

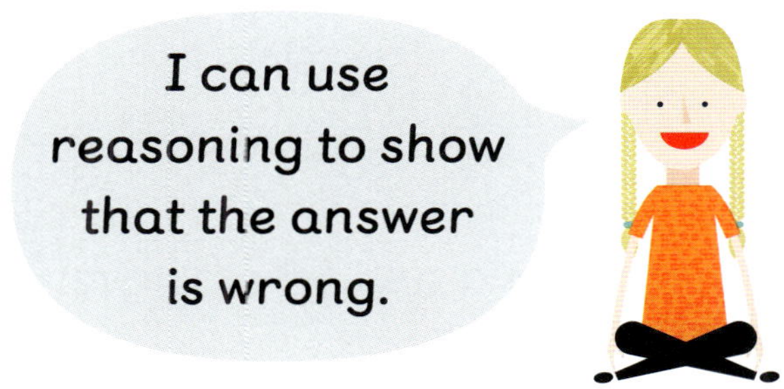

Method 2

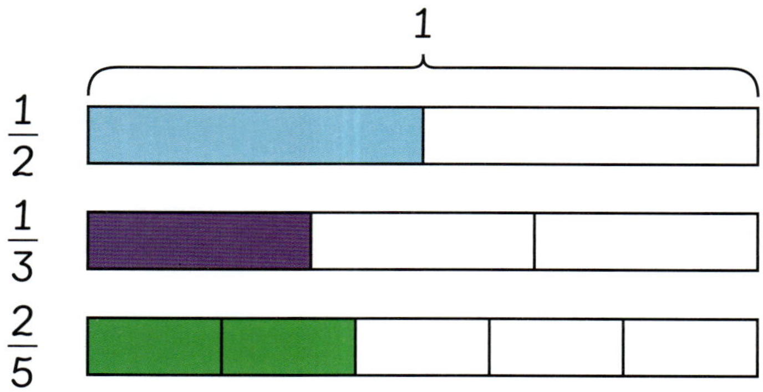

$\frac{1}{2} + \frac{1}{3}$ cannot be $\frac{2}{5}$.

2 $\frac{1}{2}$ l + $\frac{1}{3}$ l = ☐

Find a common denominator.

$\frac{1}{2} \xrightarrow{\times 3} \frac{3}{6}$ (× 3 numerator and denominator): $\frac{1}{2} = \frac{3}{6}$

$\frac{1}{3} \xrightarrow{\times 2} \frac{2}{6}$ (× 2 numerator and denominator): $\frac{1}{3} = \frac{2}{6}$

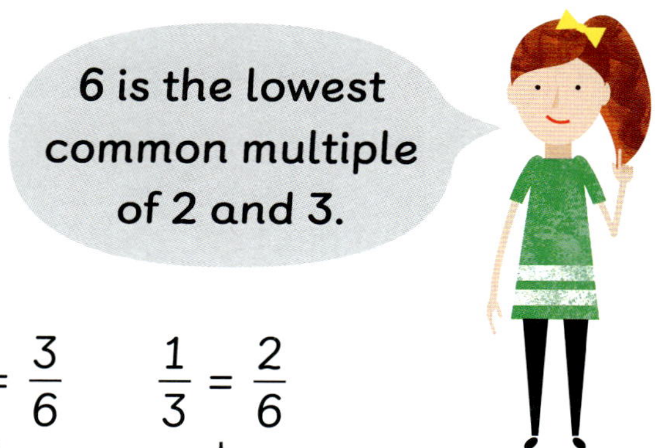

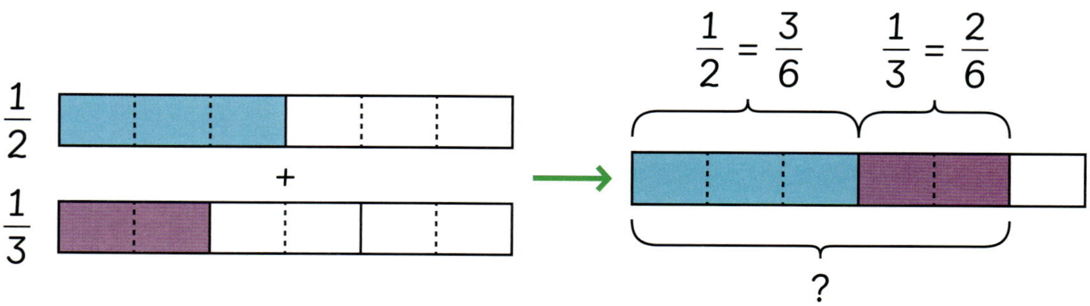

$\frac{1}{2} + \frac{1}{3} = \frac{5}{6}$

Amira and Elliott have $\frac{5}{6}$ l of juice altogether.

3 How much more juice does Amira have than Elliott?

$\frac{1}{2}l - \frac{1}{3}l = $ ☐

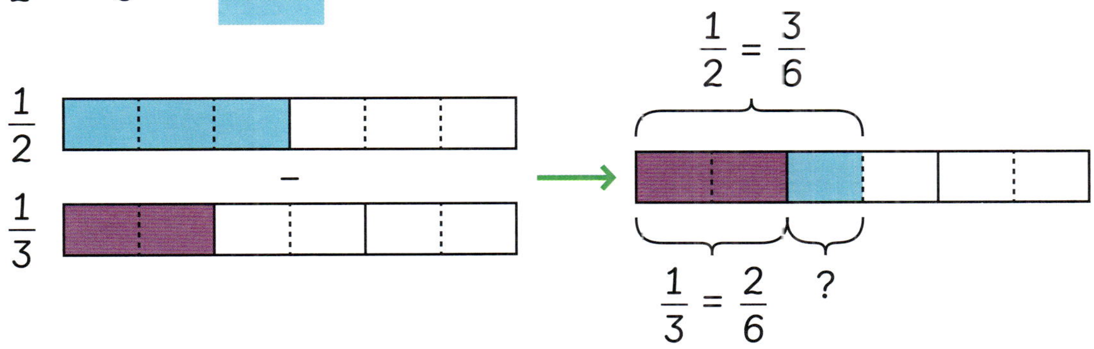

$\frac{1}{2} - \frac{1}{3} = \frac{1}{6}$

Amira has $\frac{1}{6}$ l more juice than Elliott.

Guided Practice

1 Add.

(a) $\frac{1}{2} + \frac{1}{7} = $ ☐

(b) $\frac{2}{3} + \frac{1}{5} = $ ☐

(c) $\frac{3}{10} + \frac{1}{4} = $ ☐

(d) $\frac{2}{5} + \frac{3}{7} = $ ☐

2 Subtract.

(a) $\frac{1}{2} - \frac{2}{9} = $ ☐

(b) $\frac{7}{8} - \frac{1}{6} = $ ☐

(c) $\frac{5}{6} - \frac{3}{4} = $ ☐

(d) $\frac{4}{5} - \frac{2}{3} = $ ☐

Complete Worksheet **6**: pages **71–72**

Adding and Subtracting Unlike Fractions

Lesson 7

Explore

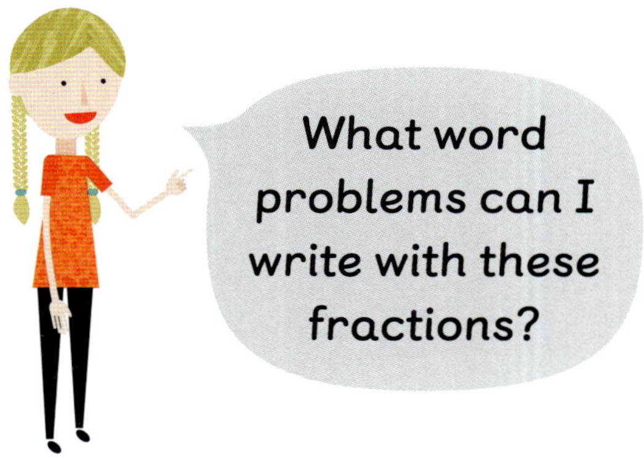

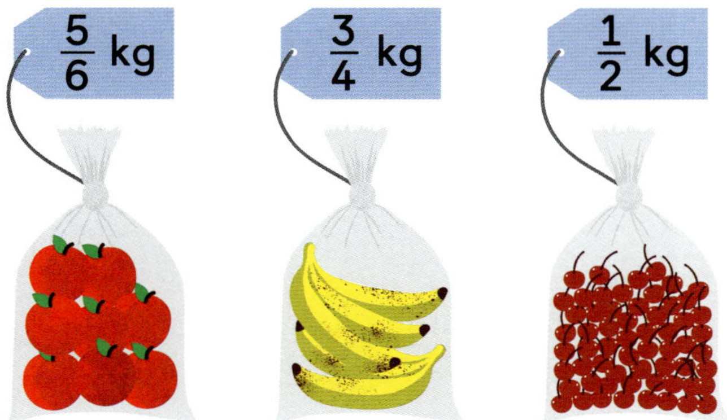

Master

1 Find a common denominator for the fractions.

$$\frac{5}{6} \overset{\times 2}{=} \frac{10}{12} \qquad \frac{3}{4} \overset{\times 3}{=} \frac{9}{12} \qquad \frac{1}{2} \overset{\times 6}{=} \frac{6}{12}$$

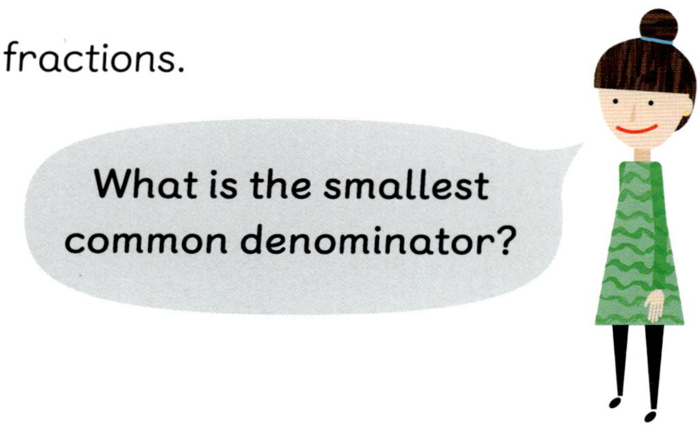

2 Which bag of fruit is the heaviest?

$\frac{5}{6}$ is the greatest fraction.

The bag of apples is the heaviest.

3 Which bag of fruit is the lightest?

$\frac{1}{2}$ is the smallest fraction.

The bag of cherries is the lightest.

4 What is the total mass of all the fruits?

$$\frac{5}{6} + \frac{3}{4} + \frac{1}{2} = \frac{10}{12} + \frac{9}{12} + \frac{6}{12}$$
$$= \frac{25}{12}$$
$$= 2\frac{1}{12}$$

The total mass is $2\frac{1}{12}$ kg.

5 How much heavier are the bananas than the cherries?

$= \frac{3}{4}$ kg

$= \frac{1}{2}$ kg

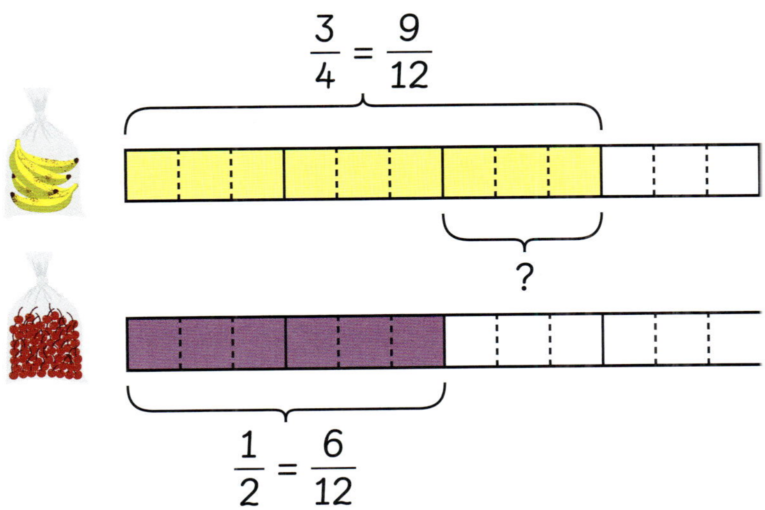

$$\frac{9}{12} - \frac{6}{12} = \frac{3}{12}$$

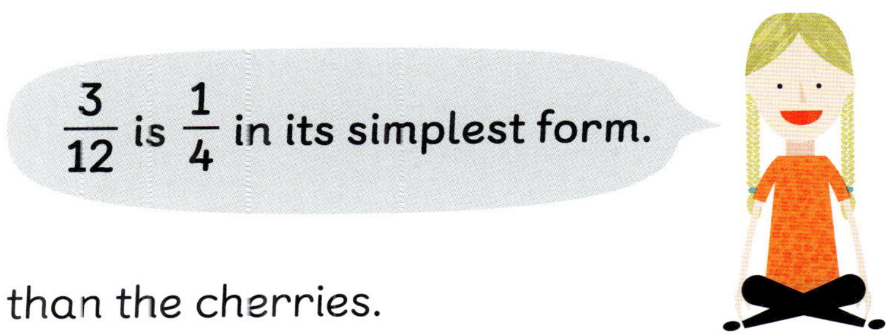

The bananas are $\frac{1}{4}$ kg heavier than the cherries.

Guided Practice

1 Find the sum.

(a) $\frac{2}{5}$ kg + $\frac{1}{3}$ kg = ☐ kg

(b) $\frac{3}{4}$ ml + $\frac{1}{10}$ ml = ☐ ml

2 Find the difference.

(a) $\frac{4}{5}$ m − $\frac{2}{3}$ m = ☐ m

(b) $\frac{2}{3}$ l − $\frac{1}{4}$ l = ☐ l

3 Solve.

(a) Emma packs a stack of books with a mass of $\frac{2}{3}$ kg in a cardboard box that has a mass of $\frac{1}{8}$ kg.
What is the total mass of the books and the cardboard box?

(b) A strip of foil is $\frac{1}{6}$ m long. A strip of paper is $\frac{3}{10}$ m longer.
Find the length of the strip of paper.

(c) Jacob has a pack of shortbread with a mass of $\frac{5}{9}$ kg and a pack of digestives with a mass of $\frac{1}{6}$ kg.
How much heavier is the pack of shortbread?

(d) A bottle holds $\frac{7}{8}$ l of syrup. Ravi pours out $\frac{5}{6}$ l of the syrup.
Find the volume of syrup left in the bottle.

Complete Worksheet **7**: pages **73–74**

Adding and Subtracting Mixed Numbers

Lesson 8

Explore

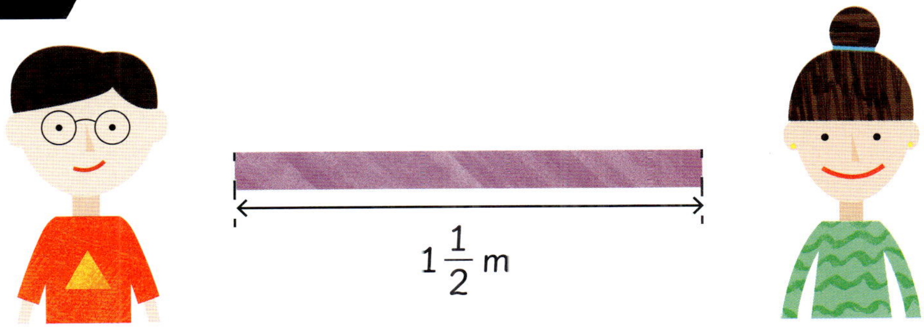

Charles and Emma share $1\frac{1}{2}$ m of ribbon.

If Charles takes $\frac{1}{3}$ m first, how much ribbon is left?

If Emma takes $\frac{3}{4}$ m first, how much ribbon is left?

Master

1 $1\frac{1}{2}\text{ m} - \frac{1}{3}\text{ m} = \square$

Method 1

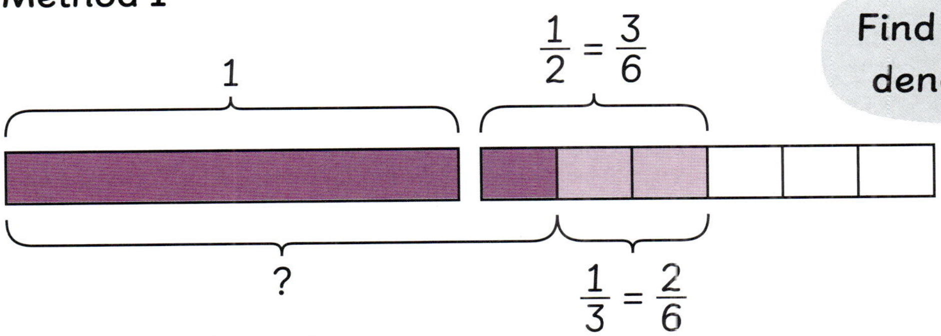

$$1\frac{1}{2} - \frac{1}{3} = 1\frac{3}{6} - \frac{2}{6}$$
$$= 1\frac{1}{6}$$

Method 2

$$1\frac{1}{2} - \frac{1}{3} = 1 - \frac{1}{3} + \frac{1}{2}$$
$$= \frac{2}{3} + \frac{1}{2}$$
$$= \frac{4}{6} + \frac{3}{6}$$
$$= \frac{7}{6}$$
$$= 1\frac{1}{6}$$

Method 3

$$1\frac{1}{2} - \frac{1}{3} = \frac{9}{6} - \frac{2}{6}$$
$$= \frac{7}{6}$$
$$= 1\frac{1}{6}$$

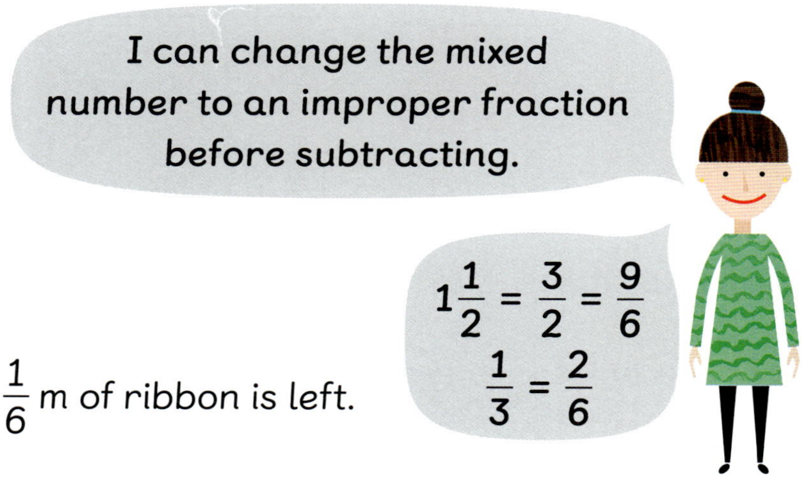

If Charles takes $\frac{1}{3}$ m first, $1\frac{1}{6}$ m of ribbon is left.

2 $1\frac{1}{2}$ m $- \frac{3}{4}$ m $=$ ▭

Method 1

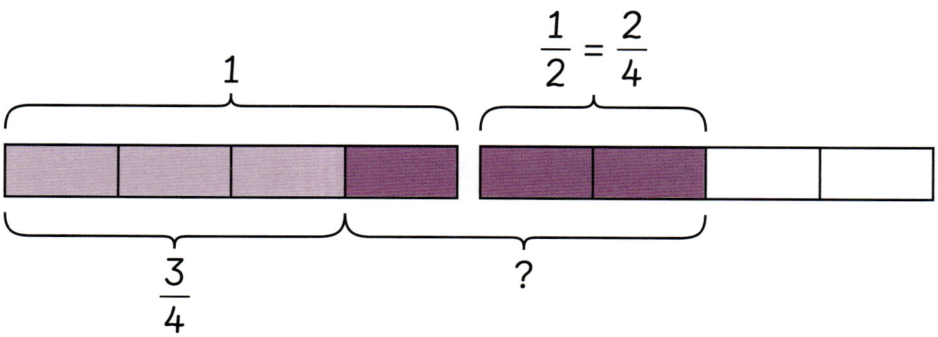

$$1\frac{1}{2} - \frac{3}{4} = \frac{1}{4} + \frac{2}{4}$$
$$= \frac{3}{4}$$

Method 2

$$1\frac{1}{2} - \frac{3}{4} = 1\frac{2}{4} - \frac{3}{4}$$
$$= \frac{6}{4} - \frac{3}{4}$$
$$= \frac{3}{4}$$

Method 3

$$1\frac{1}{2} - \frac{3}{4} = 1\frac{1}{2} - \frac{1}{2} - \frac{1}{4}$$
$$= 1 - \frac{1}{4}$$
$$= \frac{3}{4}$$

If Emma takes $\frac{3}{4}$ m first, $\frac{3}{4}$ m of ribbon is left.

3 How much ribbon is left after Charles and Emma take the length they need?

Method 1

$$1\frac{1}{2} - \frac{1}{3} = 1\frac{3}{6} - \frac{2}{6}$$
$$= 1\frac{1}{6}$$
$$1\frac{1}{6} - \frac{3}{4} = 1\frac{2}{12} - \frac{9}{12}$$
$$= \frac{14}{12} - \frac{9}{12}$$
$$= \frac{5}{12}$$

Method 2

$$\frac{1}{3} + \frac{3}{4} = \frac{4}{12} + \frac{9}{12}$$
$$= \frac{13}{12}$$
$$= 1\frac{1}{12}$$
$$1\frac{1}{2} - 1\frac{1}{12} = 1\frac{6}{12} - 1\frac{1}{12}$$
$$= \frac{5}{12}$$

$\frac{5}{12}$ m of ribbon is left over.

Guided Practice

1 Subtract or add.

(a) $1\frac{3}{4} - \frac{1}{2} =$ ☐

(b) $2\frac{3}{4} - \frac{1}{2} =$ ☐

(c) $2\frac{3}{4} + \frac{1}{2} =$ ☐

(d) $3\frac{1}{4} + \frac{1}{3} =$ ☐

2 Calculate.

(a) $1\frac{3}{4} + \frac{1}{5} =$ ☐

(b) $1\frac{1}{4} + \frac{3}{5} =$ ☐

(c) $2\frac{3}{4} - \frac{1}{5} =$ ☐

(d) $2\frac{3}{5} - \frac{1}{4} =$ ☐

3 Solve.

(a) A pack of green grapes has a mass of $1\frac{3}{4}$ kg. A pack of red grapes has a mass of $\frac{5}{6}$ kg. Find the total mass of the grapes.

(b) Holly jogged for $2\frac{1}{3}$ km on Saturday. On Sunday, she jogged $\frac{4}{5}$ km less than on Saturday.
How far did she jog on Sunday?

Complete Worksheet **8**: pages **75–76**

Adding and Subtracting Mixed Numbers

Lesson 9

Explore

Lulu is making a tropical punch. After mixing the fruit juices, she accidentally spills $2\frac{1}{4}$ l of it.

How much fruit juice does she have left?

Hawaiian Tropical Punch

1 fresh pineapple cut into chunks

1 fresh mango cut into chunks

$3\frac{2}{5}$ l pineapple juice

$1\frac{1}{2}$ l mango juice

Master

1 Find the total volume of fruit juice Lulu mixes at first.

$3\frac{2}{5}\text{ l} + 1\frac{1}{2}\text{ l} = $

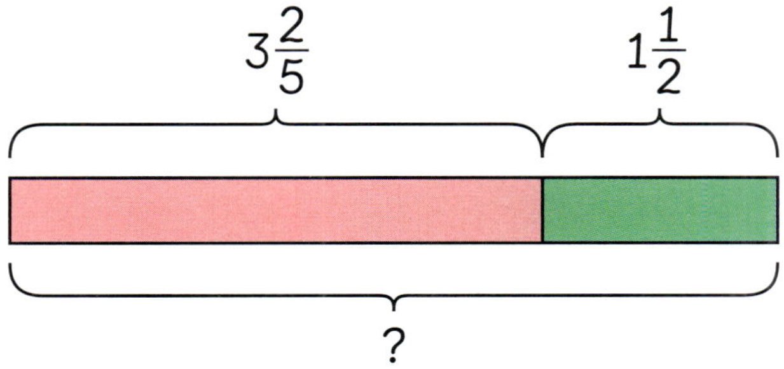

$3\frac{2}{5} = 3\frac{4}{10}$ $\qquad$ $1\frac{1}{2} = 1\frac{5}{10}$

$$3\frac{2}{5} + 1\frac{1}{2} = 3\frac{4}{10} + 1\frac{5}{10}$$
$$= 4\frac{9}{10}$$

Lulu mixes $4\frac{9}{10}$ l of fruit juice at first.

2 Find the volume of fruit juice Lulu has left.

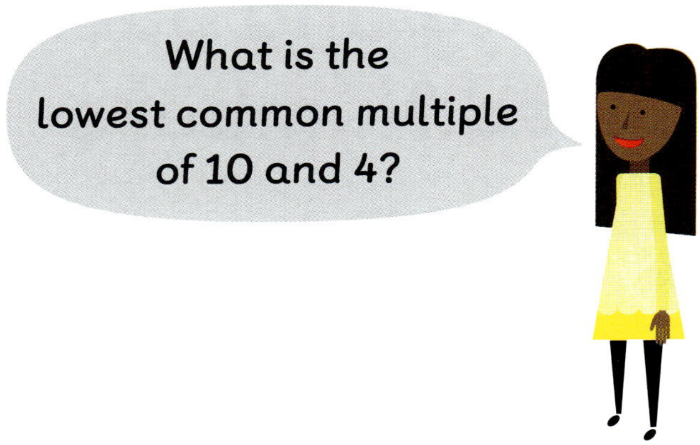

$4\frac{9}{10}\,l - 2\frac{1}{4}\,l = \square$

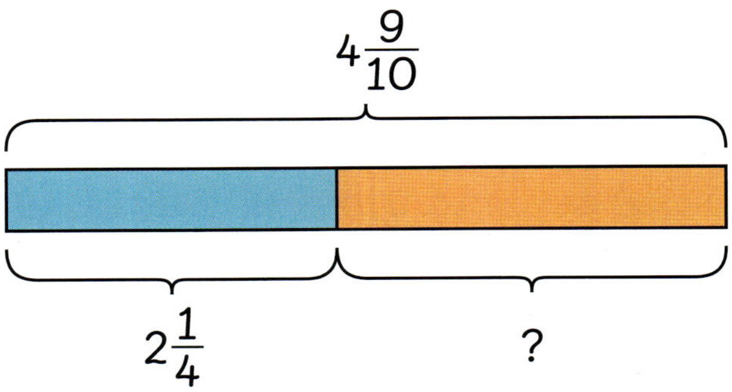

$4\frac{9}{10} = 4\frac{18}{20}$ $\qquad$ $2\frac{1}{4} = 2\frac{5}{20}$

$$4\frac{9}{10} - 2\frac{1}{4} = 4\frac{18}{20} - 2\frac{5}{20}$$

$$= 2\frac{13}{20}$$

Lulu has $2\frac{13}{20}$ l of fruit juice left.

Guided Practice

1 Add.

(a) $1\frac{3}{4} + 3\frac{5}{6} = \square$

(b) $2\frac{1}{6} + 2\frac{3}{8} = \square$

2 Subtract.

(a) $3\frac{7}{10} - 1\frac{2}{3} =$ ☐

(b) $4\frac{2}{5} - 2\frac{1}{2} =$ ☐

3 Ravi keeps track of bike rides he does with his family.
The list shows the distance they rode in each month.

(a) What is the total distance they travelled in September and October?

September	$17\frac{1}{4}$ km
October	$12\frac{2}{3}$ km
November	$21\frac{5}{6}$ km

(b) What is the difference between the total distance they travelled in the first two months, and the third month?

(c) How far did they travel in December if they travelled 70 km altogether over the four months?

Complete Worksheet **9**: pages **77–78**

Multiplying Pairs of Proper Fractions

Lesson 10

Explore

$\frac{1}{3}$ of the children in a class are on a school trip and $\frac{1}{4}$ of the remaining children are in the playground.

What fraction of the class is in the playground?

Master

1 $\frac{1}{4}$ of the children at school are in the playground.

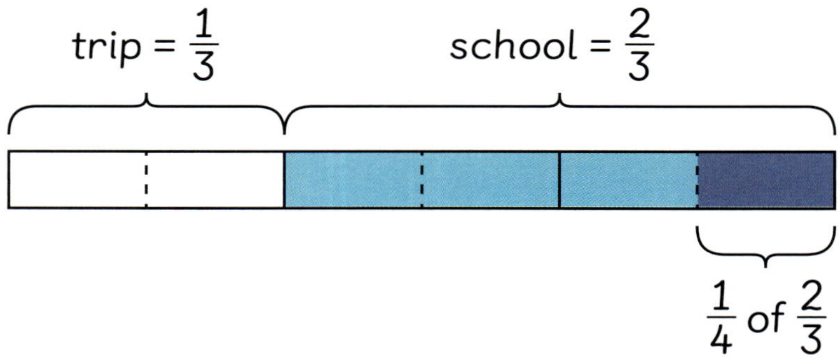

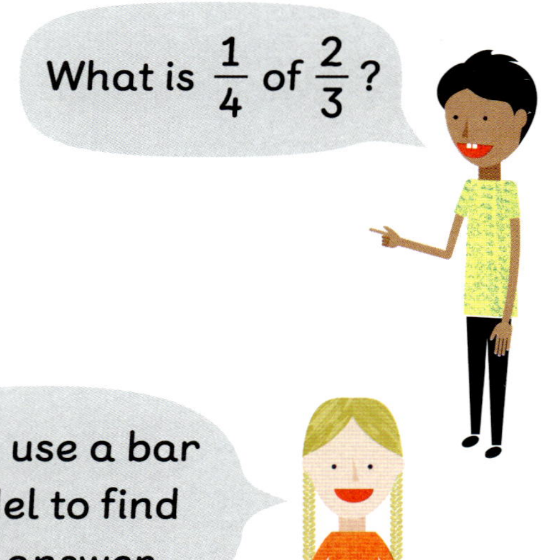

$$\frac{1}{4} \text{ of } \frac{2}{3} = \frac{1}{4} \times \frac{2}{3}$$
$$= \frac{1}{6}$$

2 $\frac{2}{3}$ of the children on the school trip took home a souvenir.

What fraction of the class had a souvenir?

$\frac{2}{3} \times \frac{1}{3} =$ ▭

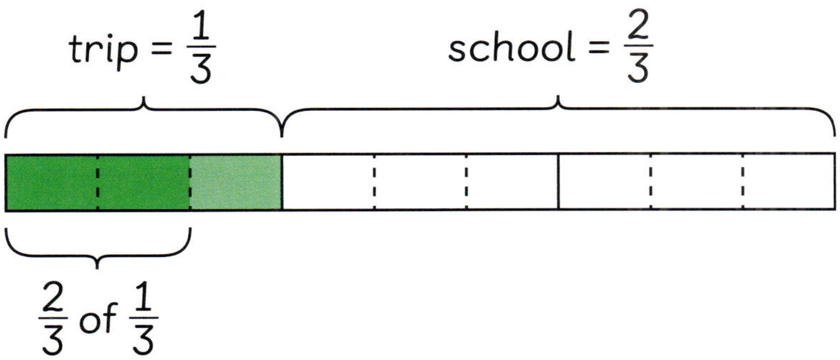

$\frac{2}{3} \times \frac{1}{3} = \frac{2}{9}$

$\frac{2}{9}$ of the class had a souvenir.

Guided Practice

1 Find the value.

(a) $\frac{1}{2}$ of $\frac{1}{5}$

(b) $\frac{1}{4}$ of $\frac{3}{5}$

2 Multiply. Give your answer in the simplest form.

(a) $\frac{1}{2} \times \frac{2}{3} =$ ▭

(b) $\frac{1}{3} \times \frac{3}{4} =$ ▭

(c) $\frac{1}{3} \times \frac{6}{7} =$ ▭

(d) $\frac{3}{4} \times \frac{8}{9} =$ ▭

3 Elliott ate $\frac{2}{5}$ of a box of chocolates and gave $\frac{5}{6}$ of the remainder to his friends. What fraction of the box of chocolates did he give to his friends?

Complete Worksheet **10**: pages **79–80**

Multiplying Pairs of Proper Fractions

Explore

Sam drinks $\frac{1}{3}$ of a cup with $\frac{1}{2}$ l of juice.

Emma drinks $\frac{1}{2}$ of a cup with $\frac{1}{3}$ l of juice.

Who drinks more?

Master

1 $\frac{1}{3} \times \frac{1}{2}$ l = ▢

= 1 l of juice

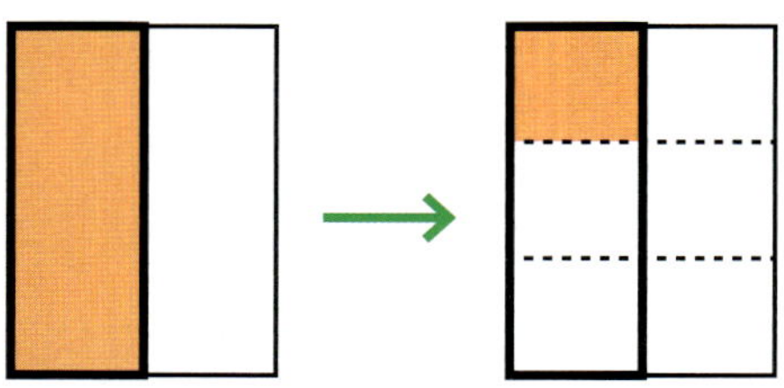

$\frac{1}{2}$ l $\qquad$ $\frac{1}{3} \times \frac{1}{2}$ l

$\frac{1}{3}$ of $\frac{1}{2}$ l is $\frac{1}{6}$ l.

Sam drinks $\frac{1}{6}$ l of juice.

2 $\frac{1}{2} \times \frac{1}{3}$ l =

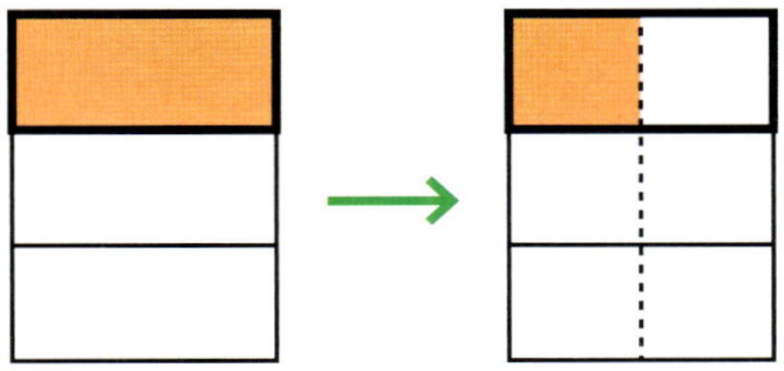

$\frac{1}{3}$ l $\frac{1}{2} \times \frac{1}{3}$ l

$\frac{1}{2} \times \frac{1}{3} = \frac{1}{6}$

$\frac{1}{2}$ of $\frac{1}{3}$ l is $\frac{1}{6}$ l.

Emma drinks $\frac{1}{6}$ l of juice.

Guided Practice

Find the value of:

1 $\frac{1}{2} \times \frac{1}{5}$

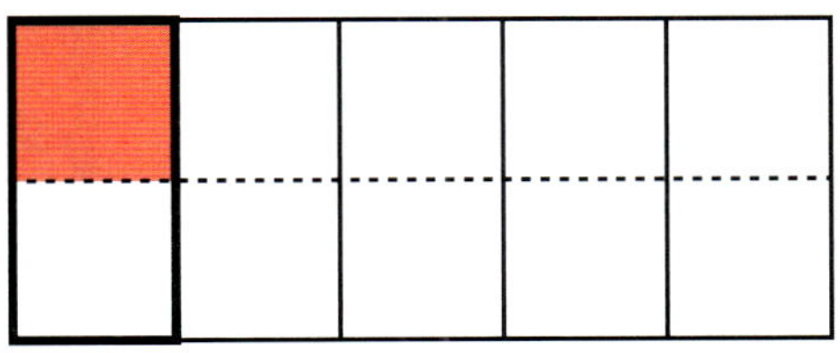

2 $\frac{1}{5} \times \frac{1}{2}$

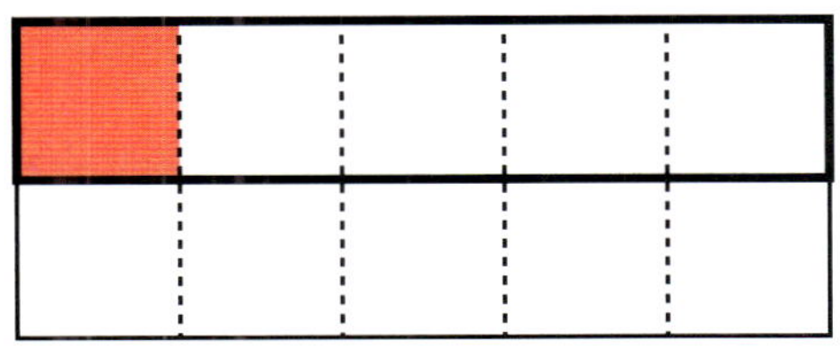

3 $\frac{1}{2} \times \frac{1}{4}$

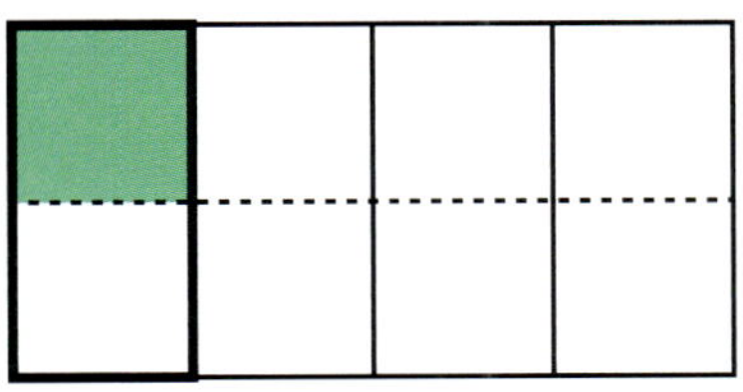

4 $\frac{1}{4} \times \frac{1}{2}$

5 $\frac{1}{3} \times \frac{1}{4}$

6 $\frac{1}{4} \times \frac{1}{3}$

Complete Worksheet **11**: pages **81–82**

Multiplying Pairs of Proper Fractions

Lesson 12

Explore

Ravi has 1 l of apple juice in a carton. He pours $\frac{2}{5}$ l of apple juice into a jug and drinks $\frac{1}{4}$ of it. How much apple juice did he drink?

Master

1 $\frac{1}{4} \times \frac{2}{5}$ l = ▭

We need to find 1 quarter of 2 fifths. I can multiply $\frac{1}{4}$ by $\frac{2}{5}$.

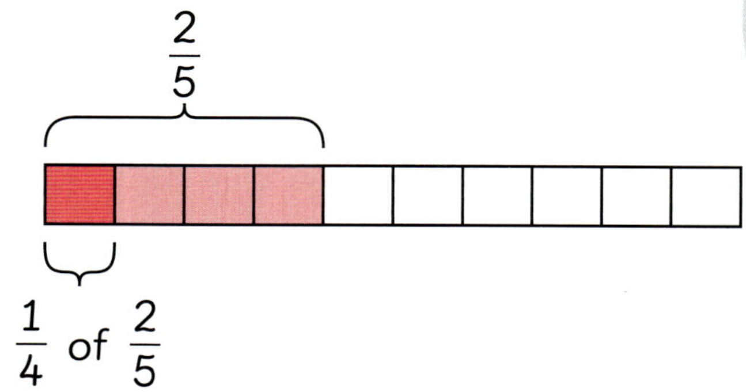

Method 1

$$\frac{1}{4} \times \frac{2}{5} = \frac{1 \times 2}{4 \times 5}$$
$$= \frac{2}{20}$$
$$= \frac{1}{10}$$

Multiply the numerator. Then multiply the denominator.

Always give the answer in the simplest form.

Method 2

$$\frac{1}{4} \times \frac{2}{5} = \frac{1}{\cancel{4}_{2}} \times \frac{\cancel{2}^{1}}{5}$$

$$= \frac{1}{2} \times \frac{1}{5}$$

$$= \frac{1}{10}$$

He drank $\frac{1}{10}$ l of apple juice.

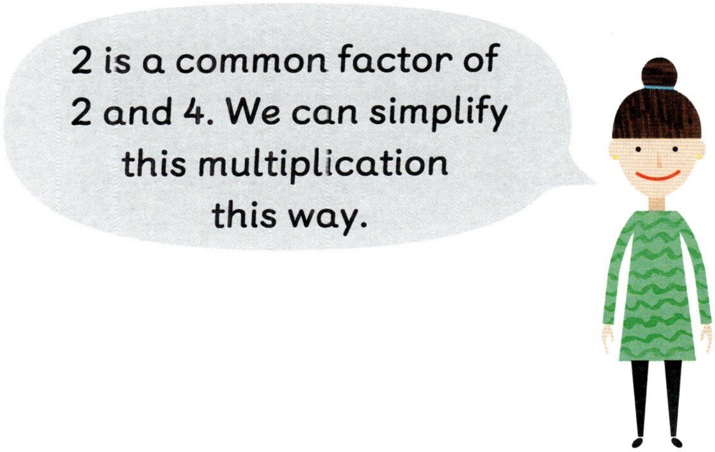

2 Ravi gives $\frac{5}{6}$ of the remaining apple juice in the carton to Sam.

How much apple juice does Sam get?

$$\frac{5}{6} \times \frac{3}{5} = \frac{\cancel{5}^{1}}{\cancel{6}_{2}} \times \frac{\cancel{3}^{1}}{\cancel{5}_{1}}$$

$$= \frac{1}{2} \times \frac{1}{1}$$

$$= \frac{1}{2}$$

Sam gets $\frac{1}{2}$ l of apple juice.

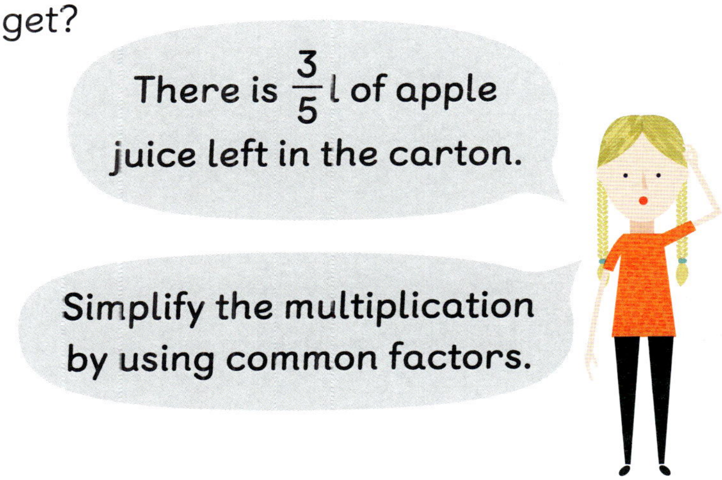

Guided Practice

Multiply. Give your answer in the simplest form.

1 $\frac{1}{4} \times \frac{2}{5} =$ ☐

2 $\frac{1}{9} \times \frac{3}{4} =$ ☐

3 $\frac{3}{8} \times \frac{2}{3} =$ ☐

4 $\frac{4}{7} \times \frac{7}{10} =$ ☐

5 $\frac{5}{18} \times \frac{9}{10} =$ ☐

6 $\frac{3}{4} \times \frac{2}{15} =$ ☐

Complete Worksheet **12**: pages **83–84**

Dividing a Fraction by a Whole Number

Lesson 13

Explore

Holly cuts a circle into thirds.
She shares 2 of the pieces between herself and Charles.
What fraction of the circle does each of them get?

Master

1 $\frac{2}{3} \div 2 = \frac{\square}{\square}$

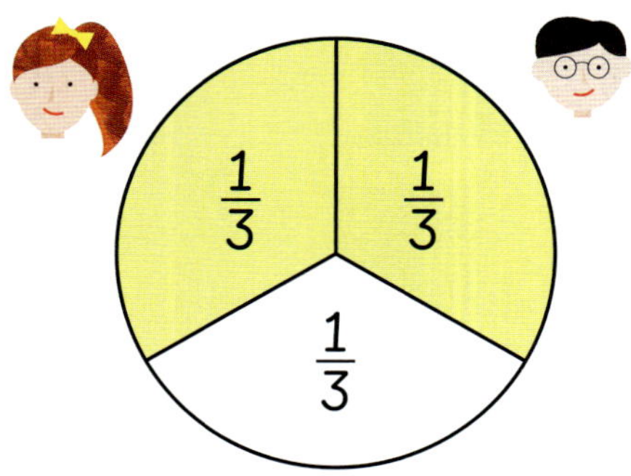

$\frac{2}{3} \div 2 = \frac{1}{3}$

2 $\frac{2}{3} \div 2 = \frac{\square}{\square}$

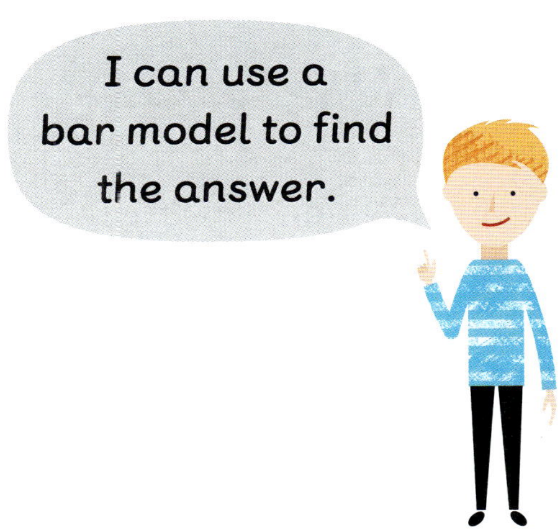

$\frac{2}{3}$

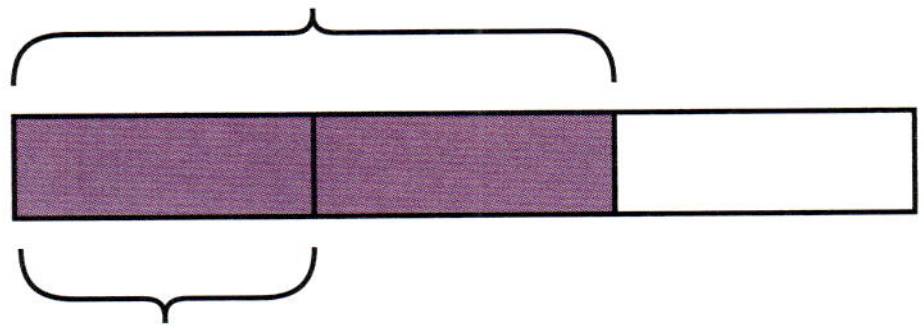

half of $\frac{2}{3}$

$\frac{2}{3} \div 2 = \frac{1}{3}$

3 $\frac{2}{3} \div 2 = \frac{1}{2}$ of $\frac{2}{3}$

$= \frac{1}{\cancel{2}_1} \times \frac{\cancel{2}^1}{3}$

$= \frac{1}{3}$

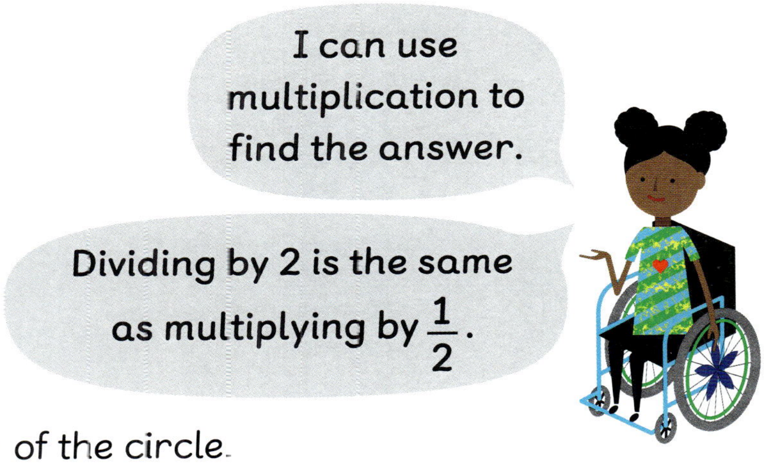

Holly and Charles each get $\frac{1}{3}$ of the circle.

4

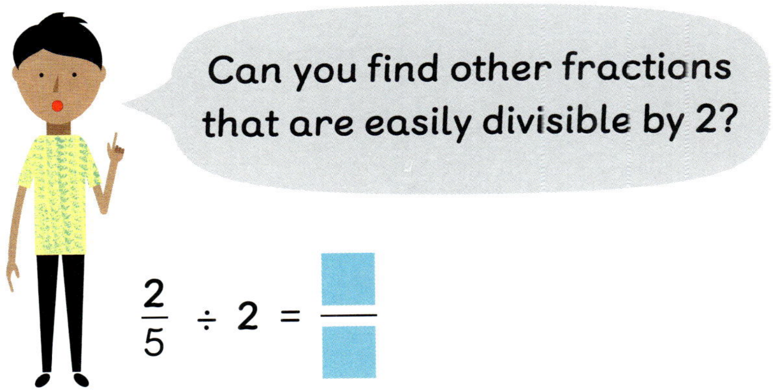

$\frac{2}{5} \div 2 = \frac{\square}{\square}$

$\frac{2}{7} \div 2 = \frac{\square}{\square}$

$\frac{2}{9} \div 2 = \frac{\square}{\square}$

Guided Practice

1 Divide.

(a) $3 \div 3 = \square$

(b) $\frac{3}{4} \div 3 = \square$

(c) $\frac{4}{5} \div 4 = \square$

2 Divide.

(a) $8 \div 2 = \square$

(b) $\frac{8}{9} \div 2 = \square$

(c) $\frac{8}{13} \div 2 = \square$

3 Divide.

(a) $\frac{15}{17} \div 3 = \square$

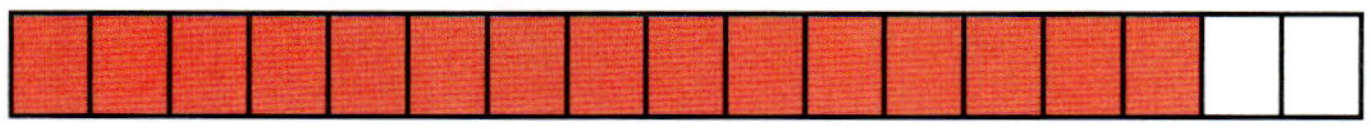

(b) $\frac{15}{17} \div 5 = \square$

(c) $\frac{18}{19} \div 2 = \square$

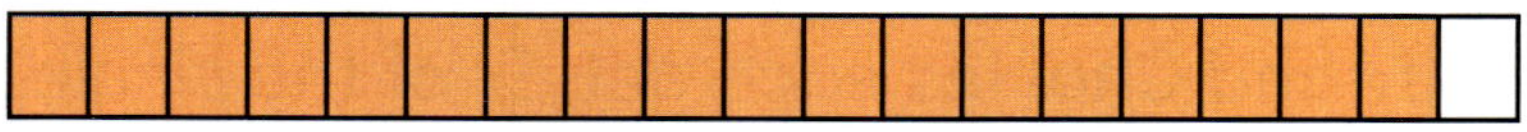

(d) $\frac{18}{19} \div 3 = \square$

(e) $\frac{18}{19} \div 6 = \square$

(f) $\frac{18}{19} \div 9 = \square$

Complete Worksheet **13**: pages **85–86**

Dividing a Fraction by a Whole Number

Lesson 14

Explore

Is Amira correct?

Master

1 $\frac{3}{4} \div 4 = \square$

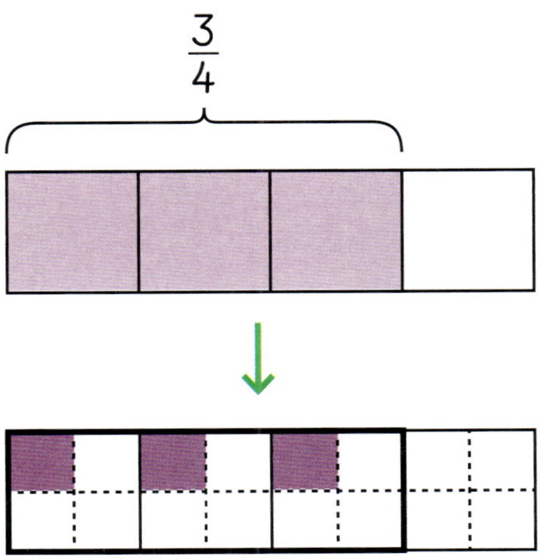

$\frac{3}{4} \div 4 = \frac{3}{16}$

2 $\frac{3}{4} \div 4 = \square$

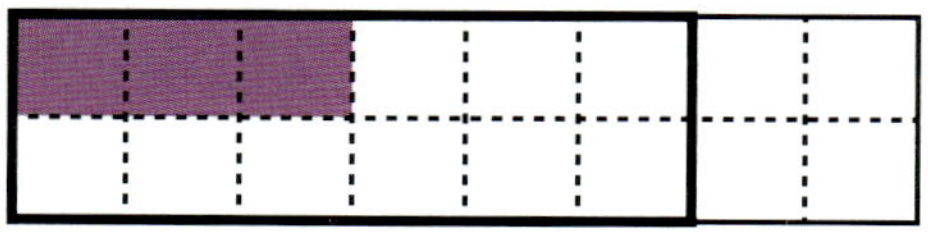

$\frac{3}{4} \div 4 = \frac{1}{4} \times \frac{3}{4} = \frac{3}{16}$

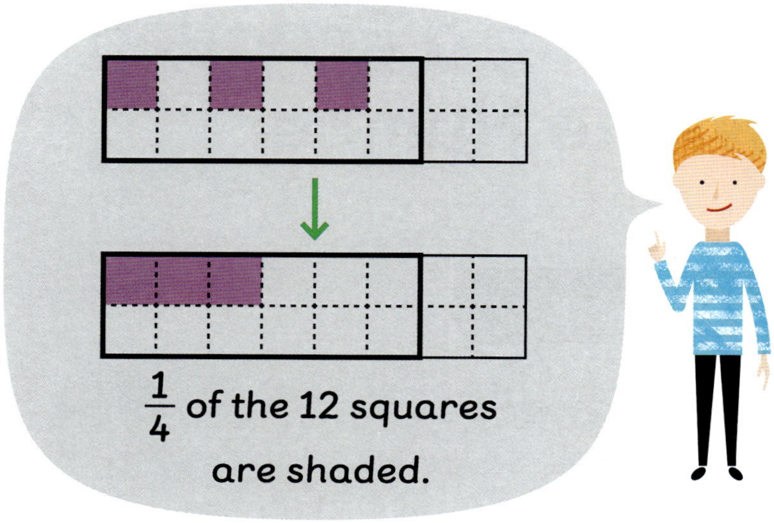

3 $\frac{3}{4} \div 4 = \frac{3}{4} \times \frac{1}{4} = \frac{3}{16}$

Amira is not correct.

Guided Practice

1 Find the quotient.

(a) $\frac{8}{9} \div 2 = \square$ (b) $\frac{8}{15} \div 4 = \square$ (c) $\frac{2}{3} \div 3 = \square$

2 Use two different methods to divide.

(a) $\frac{3}{4} \div 2 = \square$ (b) $\frac{3}{4} \div 3 = \square$ (c) $\frac{3}{4} \div 4 = \square$

3 Express the following as the product of two fractions.

(a) $\frac{3}{5} \div 2 = \frac{1}{\square} \times \frac{3}{5}$

(b) $\frac{3}{5} \div 4 = \frac{\square}{\square} \times \frac{3}{5}$

(c) $\frac{4}{5} \div 3 = \frac{\square}{\square} \times \frac{4}{5}$

(d) $\frac{4}{5} \div 7 = \frac{\square}{\square} \times \frac{4}{5}$

Complete Worksheet **14**: pages **87–88**

Dividing a Fraction by a Whole Number

Lesson 15

Explore

Elliott is dividing ingredients into 4 equal portions.
How much of each food item does he have in each portion?

Master

1 $\frac{2}{3}$ kg ÷ 4 = ☐

$\frac{2}{3}$

$\frac{4}{6}$

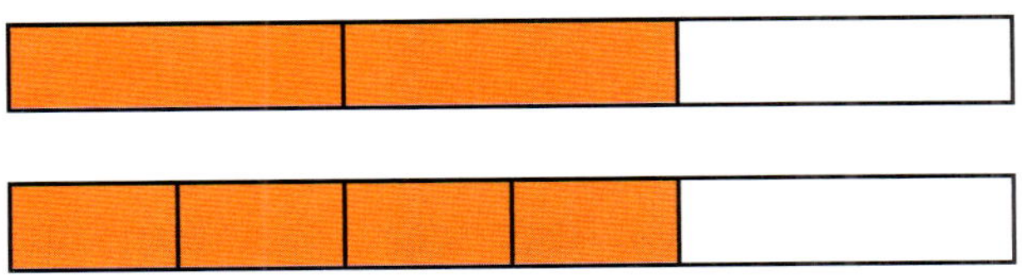

$\frac{2}{3} \div 4 = \frac{4}{6} \div 4$

$= \frac{1}{6}$

Elliott has $\frac{1}{6}$ kg of potatoes in each portion.

2 $\frac{3}{4}$ kg ÷ 4 = ☐

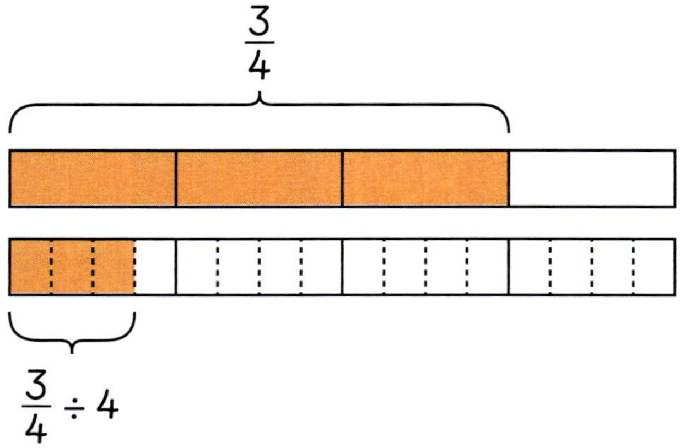

$$\frac{3}{4}\text{ kg} \div 4 = \frac{3}{4} \times \frac{1}{4}$$

$$= \frac{3}{16}$$

Elliott has $\frac{3}{16}$ kg of onions in each portion.

3 $\frac{8}{10}$ kg ÷ 4 = ☐

$$\frac{8}{10} \div 4 = \frac{1}{\not{4}_{1}} \times \frac{\not{8}^{2}}{10}$$

$$= \frac{1}{1} \times \frac{\not{2}^{☐}}{\not{10}_{☐}}$$

$$= \frac{☐}{☐}$$

Elliott has ☐ kg of peppers in each portion.

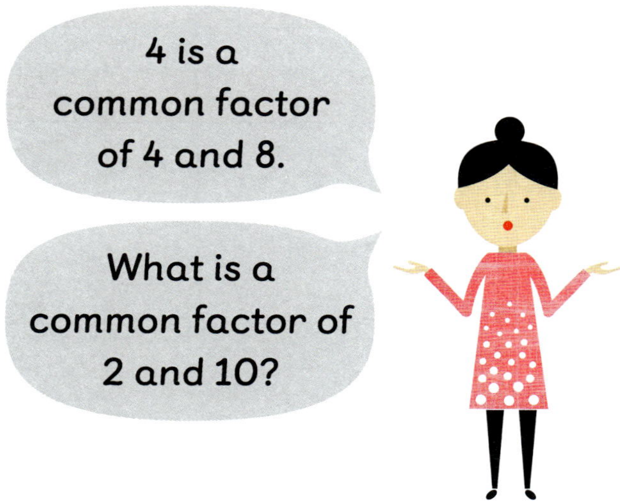

Guided Practice

1 Divide.

(a) $\frac{1}{4} \div 2 = \frac{2}{8} \div 2$

$= \square$

(b) $\frac{1}{3} \div 3 = \square \div 3$

$= \square$

(c) $\frac{2}{3} \div 3 = \square$

(d) $\frac{3}{5} \div 2 = \square$

2 Divide.

(a) $\frac{1}{4} \div 4 = \frac{1}{4} \times \frac{1}{4}$

$= \square$

(b) $\frac{2}{3} \div 5 = \frac{2}{3} \times \square$

$= \square$

(c) $\frac{4}{5} \div 3 = \square$

(d) $\frac{7}{9} \div 2 = \square$

3 $\frac{7}{8}$ l of liquid plant food is poured equally into 9 plant pots.
What is the volume of liquid plant food poured into each pot?

Complete Worksheet **15**: pages **89–90**

Mind Challenge

Sam and Elliott have £81 altogether.

Sam spends $\frac{3}{7}$ of his money on tickets and snacks at the cinema.

Elliott spends £4 of his money on his lunch.
After spending the money, the boys are left with the same amount each.
How much money did each of them have to start with?

My Maths Journal

Write a word problem that requires both of the following calculations.

$\frac{1}{4} \times \frac{4}{5}$ $\frac{1}{5} \div 2$

Review

I can...

- ☐ simplify fractions using common factors.
- ☐ compare and order fractions and mixed numbers.
- ☐ add and subtract fractions and mixed numbers.
- ☐ multiply proper fractions.
- ☐ divide proper fractions by whole numbers.

When do we use decimal numbers?

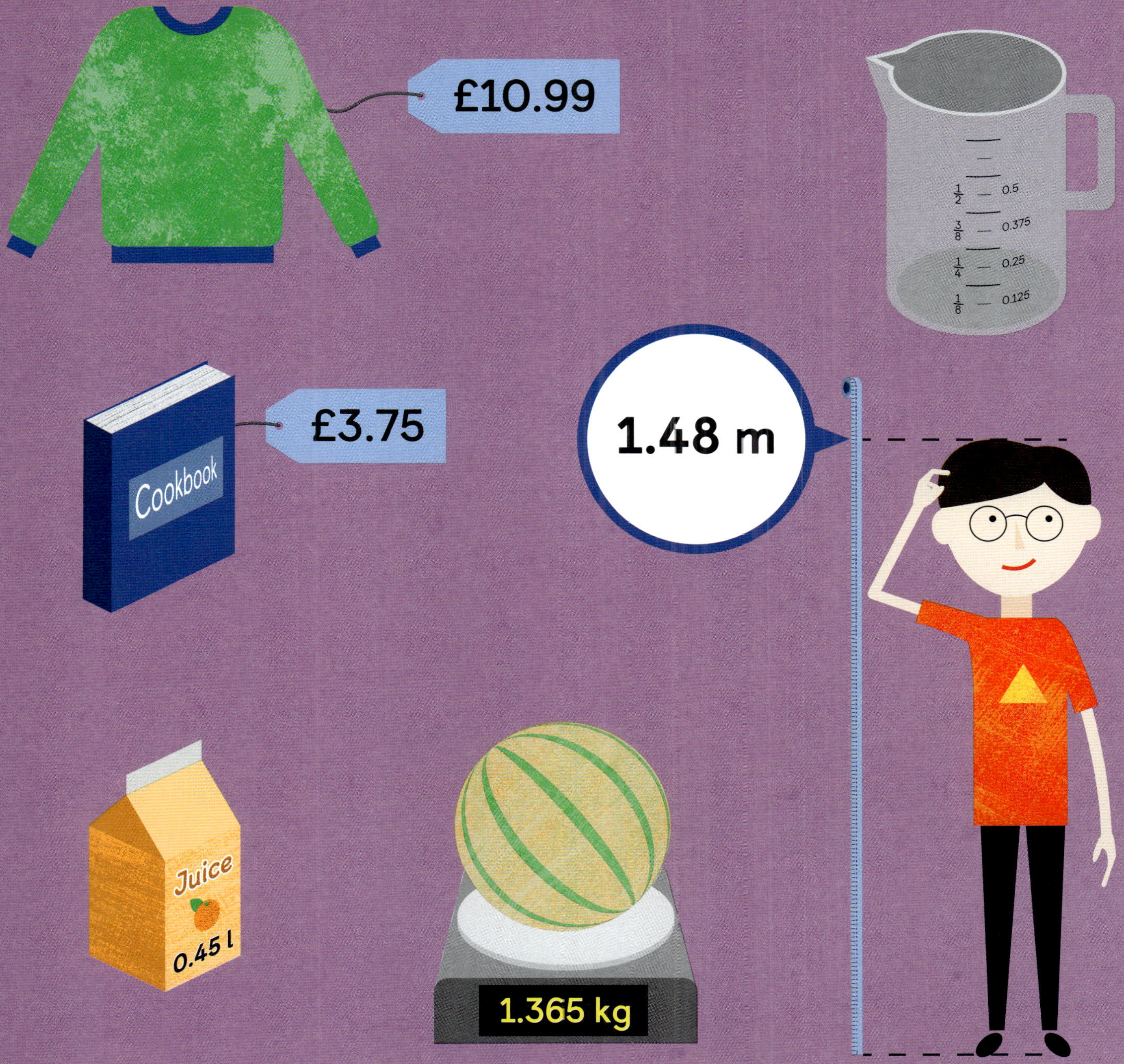

Chapter 4

Decimals

Reading and Writing Decimals

Lesson 1

Explore

Holly made 2.134 in this way:

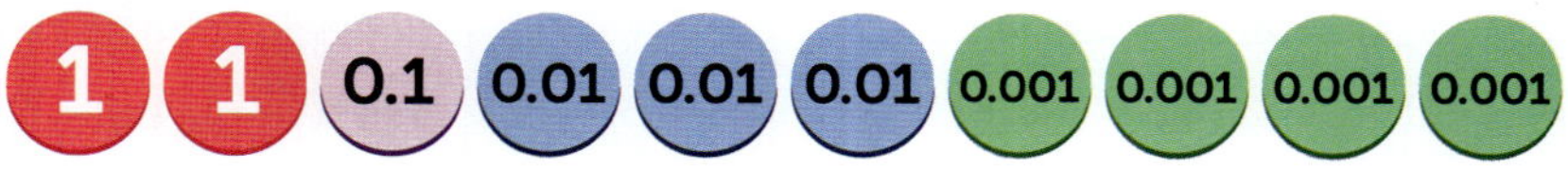

Elliott made 2.134 in this way:

What do , , 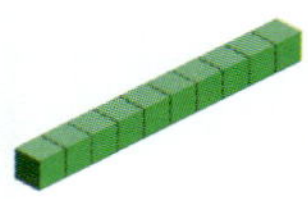, and  represent?

Master

1 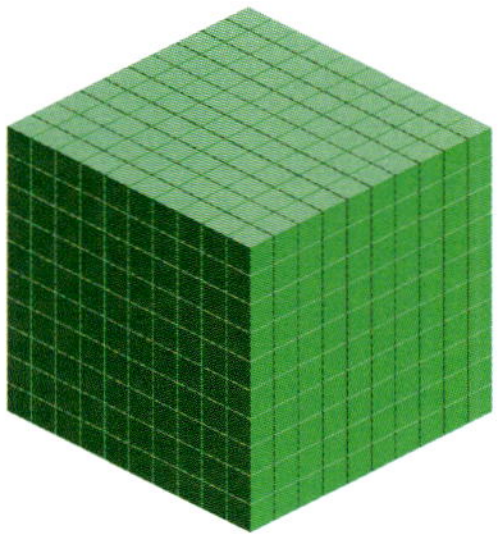 = 1

There are 10  in 1 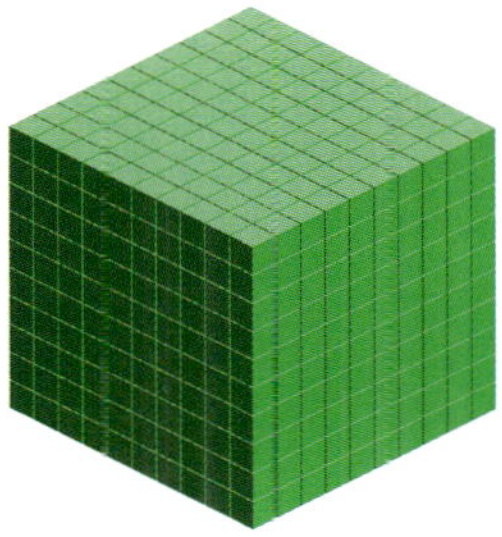.

Each 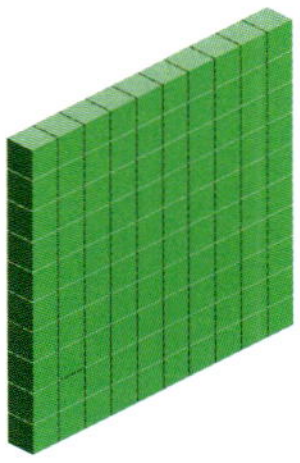 is 1 tenth.

10 tenths is equal to 1.

We write 1 tenth as $\frac{1}{10}$ or 0.1.  = 0.1

100 hundredths is equal to 1.

2 There are 100  in 1 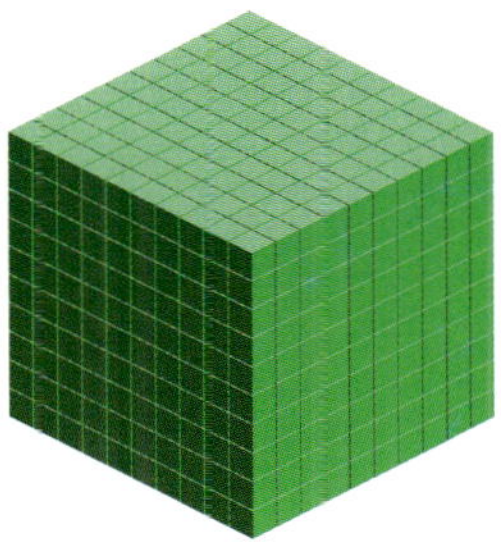.

Each 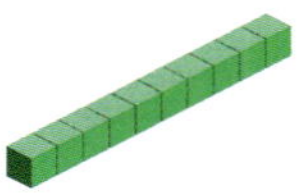 is 1 hundredth.

We write 1 hundredth as $\frac{1}{100}$ or 0.01. 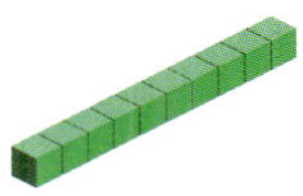 = 0.01

3 There are 1000 in 1 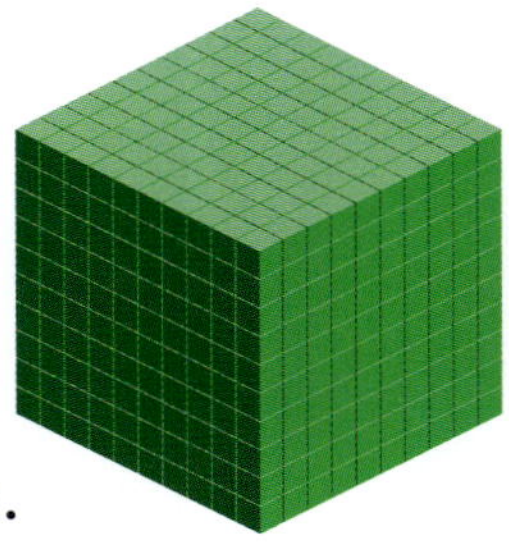 .

Each is 1 thousandth.

We write 1 thousandth as $\frac{1}{1000}$ or 0.001. = 0.001

4 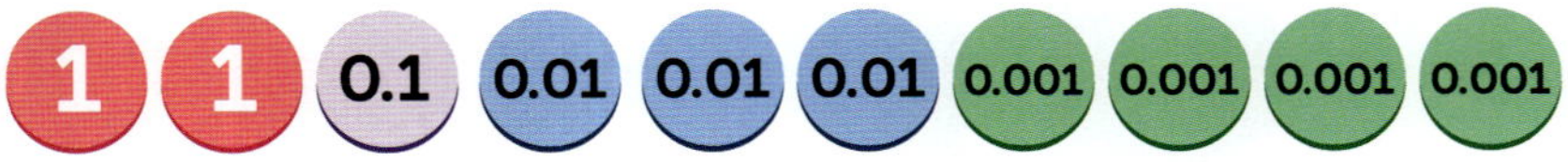

2	.	1	3	4

In 2.134, the digit 2 is in the **ones** place and stands for **2 ones**. It has a value of 2.

In 2.134 the digit 1 is in the **tenths** place and stands for **1 tenth**. It has a value of 0.1.

In 2.134 the digit 3 is in the **hundredths** place and stands for **3 hundredths**. It has a value of 0.03.

In 2.134 the digit 4 is in the **thousandths** place and stands for **4 thousandths**. It has a value of ______ .

5

0	.	3	4

0	.	7	6	1

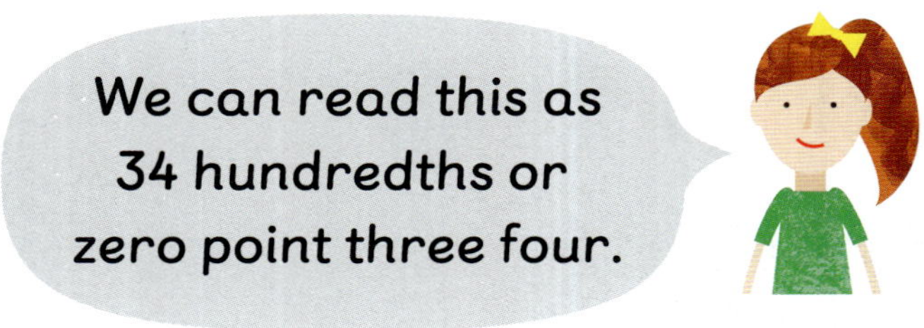

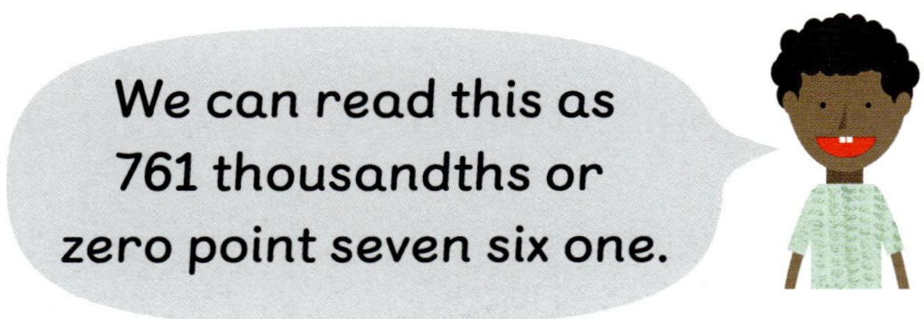

Guided Practice

1 Write the decimal numbers shown.

(a) 

(b) 

2 What is the value of 5 in each of the following numbers?

(a) 3.54

(b) 5.013

(c) 0.952

(d) 0.005

3 Use all the digits 0, 2, 4 and 6 and a . once to make a number between 2 and 3.

Complete Worksheet **1**: pages **103–104**

Dividing Whole Numbers by Multiples of 10

Lesson 2

Explore

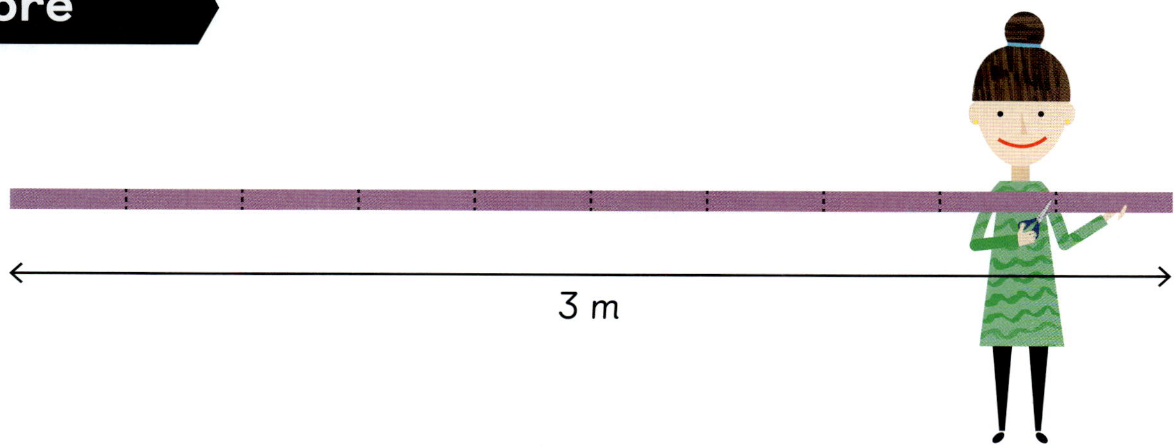

A 3 m length of ribbon is cut into 10 equal pieces.
How long is each piece?
Give your answer in metres.

Master

1 $3 \text{ m} \div 10 =$ ▭

$3 \div 10 = 30 \text{ tenths} \div 10$
$= 3 \text{ tenths}$
$= 0.3$

$3 \text{ m} \div 10 = 0.3 \text{ m}$

Each piece is 0.3 m.

2 What if we divided it into 100 pieces?

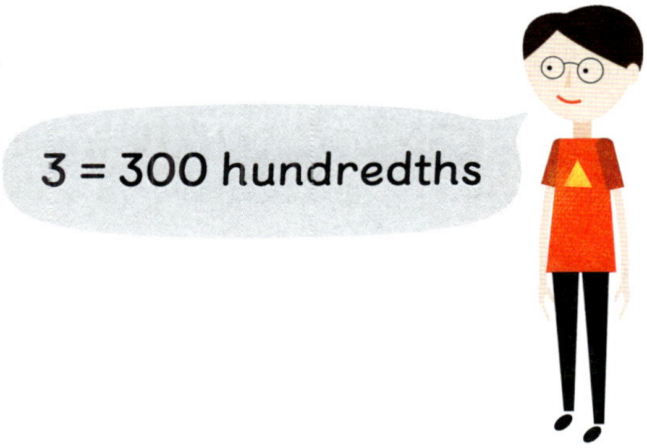

3 m ÷ 100 =

3 ÷ 100 = 300 hundredths ÷ 100
= 3 hundredths
= 0.03

3 m ÷ 100 = 0.03 m
Each piece would be 0.03 m.

3 What if we divided it into 1000 pieces?

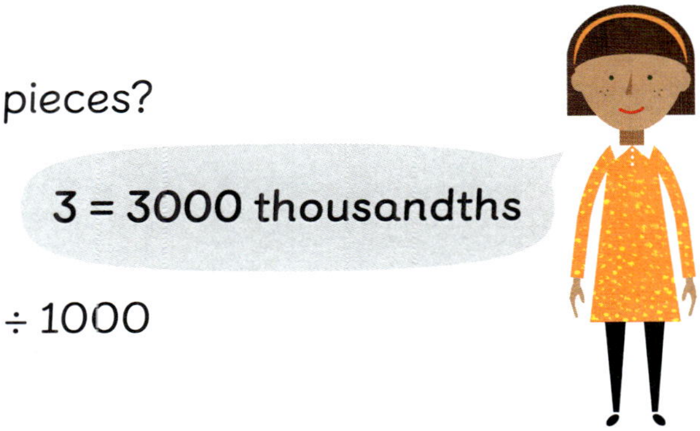

3 m ÷ 1000 =

3 ÷ 1000 = 3000 thousandths ÷ 1000
= 3 thousandths
= 0.003

3 m ÷ 1000 = 0.003 m
Each piece would be 0.003 m.

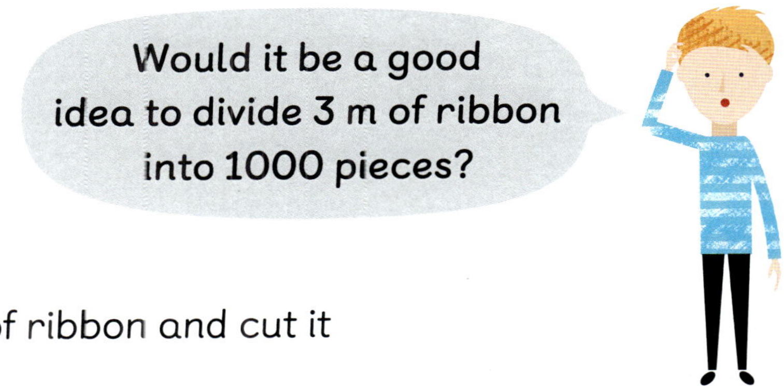

4 What if we started with 35 m of ribbon and cut it into 10 equal pieces?

35 m ÷ 10 =

35 ÷ 10 = 350 tenths ÷ 10
= 35 tenths
= 3.5

35 m ÷ 10 = 3.5 m
Each piece would be 3.5 m.

5 What if we cut the 35 m of ribbon into 100 or 1000 equal pieces?

35 m ÷ 100 = ______

35 ÷ 100 = 3500 hundredths ÷ 100
= 35 hundredths
= 0.35

35 m ÷ 100 = 0.35 m
Each piece would be 0.35 m.

35 m ÷ 1000 = ______
35 ÷ 1000 = 35 000 thousandths ÷ 1000
= 35 thousandths
= 0.035

35 m ÷ 1000 = 0.035 m
Each piece would be 0.035 m.

Guided Practice

1 Write the answer of each as a decimal.

(a) 5 ÷ 10 = ______

(b) 5 ÷ 100 = ______

(c) 5 ÷ 1000 = ______

(d) 45 ÷ 10 = ______

(e) 45 ÷ 100 = ______

(f) 45 ÷ 1000 = ______

(g) 123 ÷ 10 = ______

(h) 123 ÷ 100 = ______

(i) 123 ÷ 1000 = ______

2 Fill in the missing numbers.

(a) 321 ÷ ______ = 32.1

(b) ______ ÷ 100 = 3.21

(c) 901 ÷ 10 = ______

(d) 901 ÷ ______ = 0.901

Complete Worksheet **2**: pages **105–106**

Dividing Whole Numbers

Lesson 3

Explore

Hannah uses 5 kg of sand to make 8 identical sandcastles.
What is the mass of sand in each sandcastle?

Master

1 5 kg ÷ 8 =

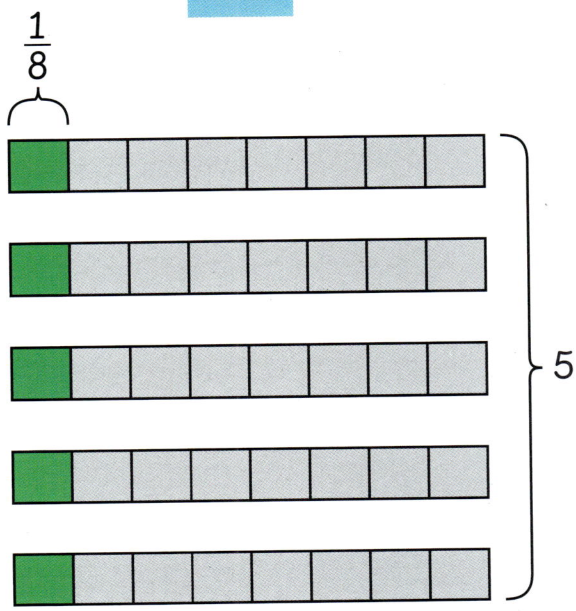

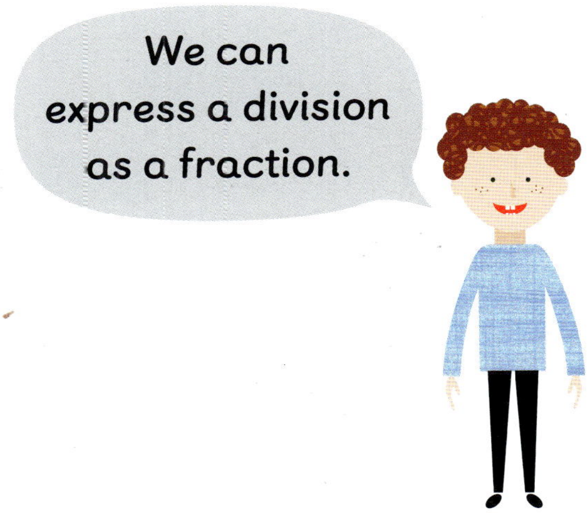

$5 \text{ kg} \div 8 = \frac{5}{8} \text{ kg}$

The mass of sand in each sandcastle is $\frac{5}{8}$ kg.

2 5 kg ÷ 8 = ______

```
        ___ ___ ___
8 ) 5  0  0  0
  − 4  8  0  0  →  8 × 600
       2  0  0
  −    1  6  0  →  8 ×  20
          4  0
  −       4  0  →  8 ×   5
             0
```

5000 thousandths ÷ 8 = 625 thousandths
625 thousandths = 0.625

$\frac{5}{8} = 0.625$

5 kg ÷ 8 = 0.625 kg
The mass of sand in each sandcastle is 0.625 kg.

Guided Practice

1 Divide.

(a) 1 ÷ 2 = ______

10 tenths ÷ 2 = ______ tenths

(b) 1 ÷ 4 = ______

100 hundredths ÷ 4 = ______ hundredths

(c) 1 ÷ 8 = ______

1000 thousandths ÷ 8 = ______ thousandths

2 Divide.

(a) $3 \div 5 =$ ____

30 tenths $\div 5 =$ ____ tenths

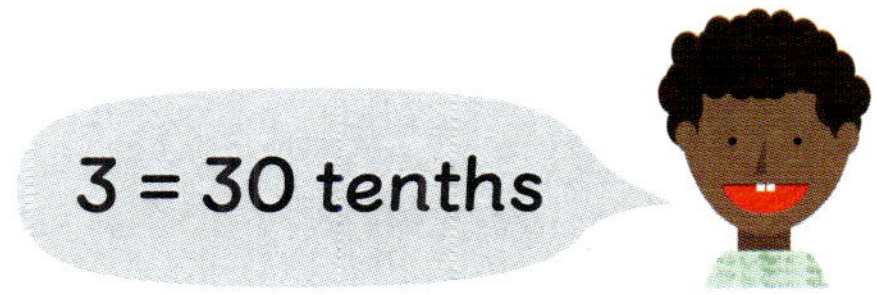

(b) $2 \div 8 =$ ____

200 hundredths $\div 8 =$ ____ hundredths

(c) $2 \div 5 =$ ____

(d) $3 \div 4 =$ ____

(e) $3 \div 8 =$ ____

(f) $7 \div 8 =$ ____

3 Sam's mum uses 4 l of paint to paint 5 fence panels.
She uses the same volume of paint on each fence panel.
What is the volume of paint she uses on a single panel?

Complete Worksheet **3**: pages **107–108**

Writing Fractions as Decimals

Lesson 4

Explore

Ravi draws a map showing some places near his house. The drawing does not accurately show how far each of these places are from his house.

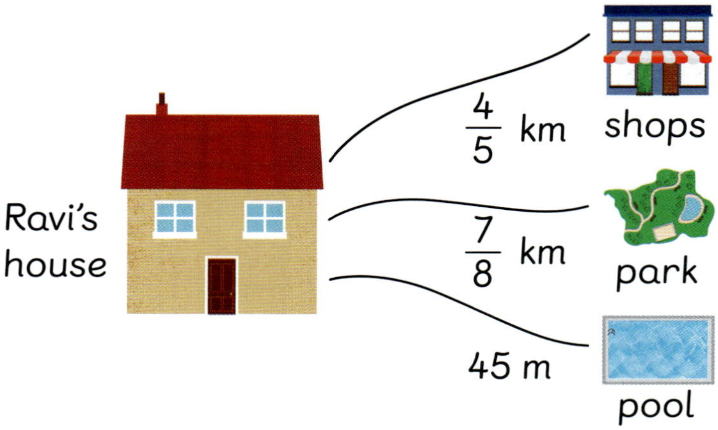

How can Ravi compare the distances accurately?
Which of the places shown is furthest from his house?

Master

1 Ravi writes the distances as decimal numbers.

$$\frac{4}{5}\text{ km} = \frac{8}{10}\text{ km}$$
$$= 0.8\text{ km}$$

2 $\frac{7}{8}$ km = 7 km ÷ 8

7 km ÷ 8 = $\square$

Method 1

$$
\begin{array}{r@{\;}r l}
 & 875 & \\
8\,\big) & 7000 & \\
- & 6400 & \longrightarrow 8 \times 800 \\
\hline
 & 600 & \\
- & 560 & \longrightarrow 8 \times 70 \\
\hline
 & 40 & \\
- & 40 & \longrightarrow 8 \times 5 \\
\hline
 & 0 &
\end{array}
$$

$\frac{7}{8}$ = 875 thousandths = 0.875

$\frac{7}{8}$ km = 0.875 km

Method 2

$\frac{7}{8}$ km = $\square$

$$\frac{7}{8} \overset{\times 125}{=} \frac{\square}{1000} \quad (\times 125)$$

3 45 m = $\square$

1000 m = 1 km

45 m = $\frac{45}{1000}$ km

= 0.045 km

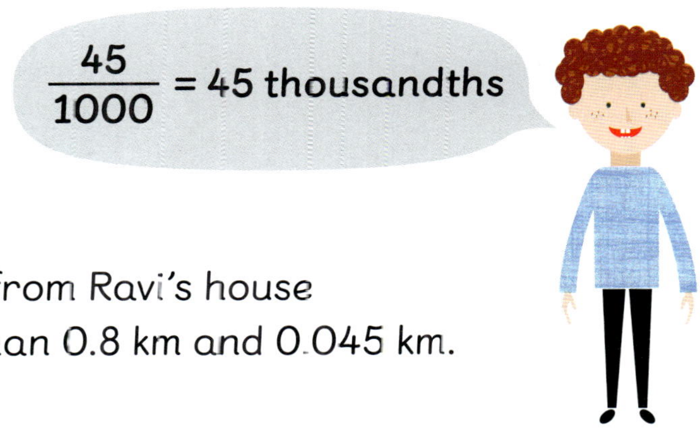

The park is the furthest from Ravi's house as 0.875 km is greater than 0.8 km and 0.045 km.

Guided Practice

1 Write each fraction as a decimal.

(a) $\frac{1}{4} = \square$

(b) $\frac{1}{8} = \square$

(c) $\frac{3}{8} = \square$

(d) $\frac{1}{2} = \square$

(e) $\frac{3}{5} = \square$

(f) $\frac{6}{8} = \square$

2 Place $\frac{1}{4}$, $\frac{1}{8}$, $\frac{3}{8}$, $\frac{1}{2}$, $\frac{3}{5}$ and $\frac{6}{8}$ in their approximate positions on the number line.

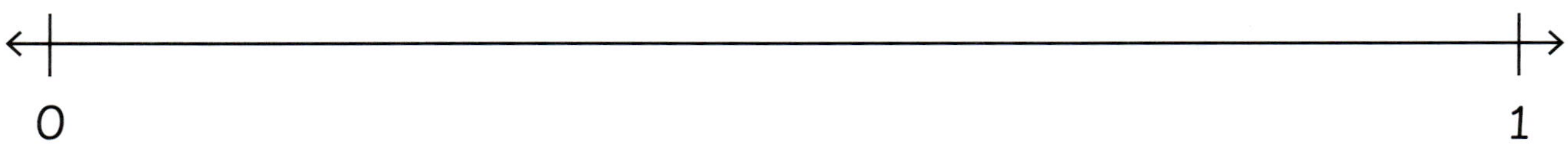

3 Write each decimal as a fraction in its simplest form.

(a) $0.2 = \square$

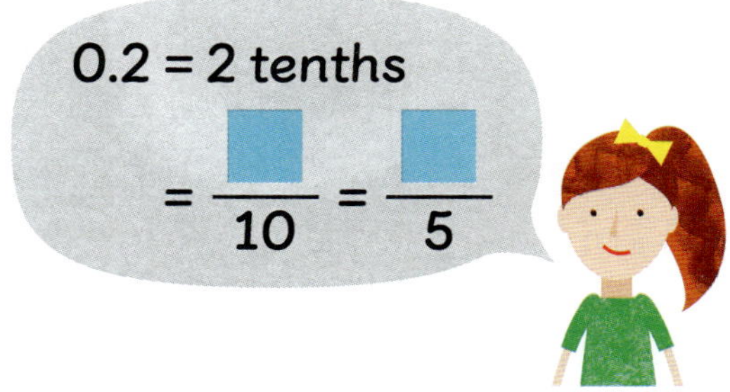

(b) $0.32 = \square$

(c) $0.125 = \square$

Complete Worksheet **4**: pages **109–110**

Writing Fractions as Decimals

Lesson 5

Explore

How can the volumes of water above be shown using decimal numbers?

Master

1 $\frac{1}{4}$ l = ______

1 = 100 hundredths

$\frac{1}{4}$ = 1 ÷ 4
= 100 hundredths ÷ 4
= 25 hundredths
= 0.25

$\frac{1}{4}$ l = 0.25 l

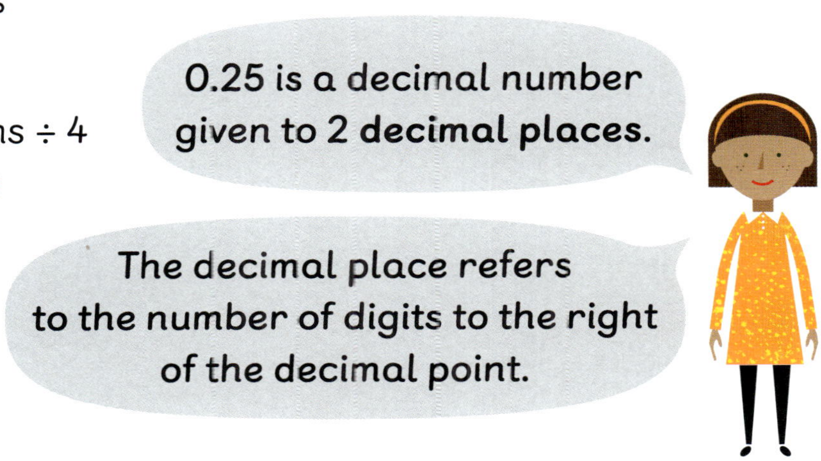

2 $\frac{1}{8}$ l = ▭

1 = 1000 thousandths

$\frac{1}{8} = 1 \div 8$

$= 1000$ thousandths $\div 8$

$= 125$ thousandths

$= 0.125$

$\frac{1}{8}$ l = 0.125 l

$$
\begin{array}{r|rrrrl}
 & & \square & \square & \square & \\
8 & 1 & 0 & 0 & 0 & \\
- & & 8 & 0 & 0 & \rightarrow 8 \times 100 \\
\hline
 & & 2 & 0 & 0 & \\
- & & 1 & 6 & 0 & \rightarrow 8 \times 20 \\
\hline
 & & & 4 & 0 & \\
- & & & 4 & 0 & \rightarrow 8 \times 5 \\
\hline
 & & & & 0 &
\end{array}
$$

3 $\frac{1}{3}$ l = ▭

$$
\begin{array}{r|rrrrl}
 & & 3 & 3 & 3 & \\
3 & 1 & 0 & 0 & 0 & \\
- & & 9 & 0 & 0 & \rightarrow 3 \times 300 \\
\hline
 & & 1 & 0 & 0 & \\
- & & & 9 & 0 & \rightarrow 3 \times 30 \\
\hline
 & & & 1 & 0 & \\
- & & & & 9 & \rightarrow 3 \times 3 \\
\hline
 & & & & 1 &
\end{array}
$$

$\frac{1}{3} = 1 \div 3$

$= 1000$ thousandths $\div 3$

≈ 333 thousandths

$= 0.333$

$\frac{1}{3}$ l ≈ 0.333 l

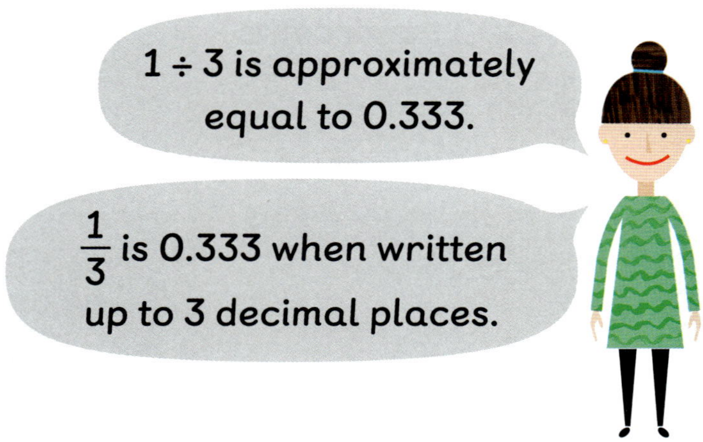

Activity Time

Partner work

Complete the table with these unit fractions.

$\frac{1}{2}$ $\frac{1}{3}$ $\frac{1}{4}$ $\frac{1}{5}$ $\frac{1}{6}$ $\frac{1}{7}$ $\frac{1}{8}$ $\frac{1}{9}$ $\frac{1}{10}$

1 decimal place	2 decimal places	3 decimal places	4 or more decimal places

Guided Practice

Express each fraction as a decimal number. Write your answers up to 3 decimal places if they are not exact.

1. $\frac{1}{10} =$ ______
2. $\frac{3}{4} =$ ______
3. $\frac{3}{8} =$ ______
4. $\frac{1}{6} =$ ______
5. $\frac{4}{7} =$ ______
6. $\frac{8}{9} =$ ______

Complete Worksheet **5**: pages **111–112**

Multiplying Decimals without Renaming

Lesson 6

Explore

Charles' dad wants to make soup using the recipe shown.
He wants to double the recipe.
For the first three items on the list, what volumes should he use?

Vegetable soup

0.4 l	coconut milk
0.34 l	water
1.3 l	vegetable stock
5	onions
3	carrots

Master

1 $2 \times 0.4 \text{ l} =$ ______

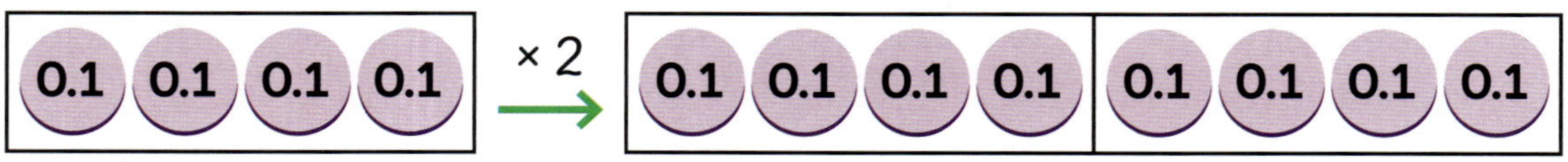

$2 \times 0.4 = 2 \times 4$ tenths
= 8 tenths
= 0.8

Charles' dad should use 0.8 l of coconut milk.

2 $2 \times 0.34\ l =$

Method 1

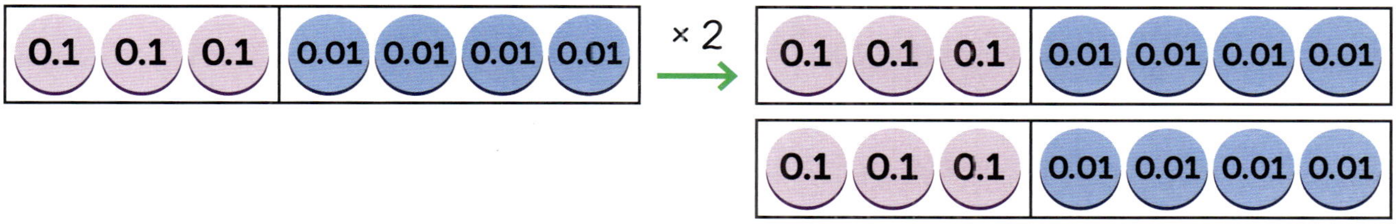

$2 \times 0.3 = 0.6$
$2 \times 0.04 = 0.08$
$2 \times 0.34 = 0.6 + 0.08$
$= 0.68$

Method 2

$2 \times 0.34 = 2 \times 34$ hundredths
$= 68$ hundredths
$= 0.68$

Charles' dad should use 0.68 l of water.

3 $2 \times 1.3\ l =$

Method 1

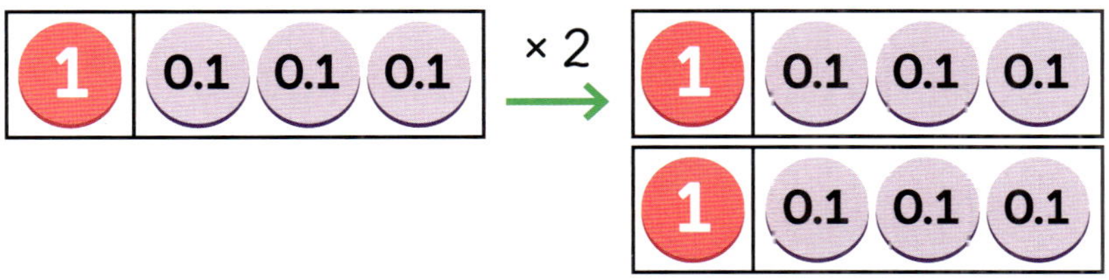

$2 \times 1 = 2$
$2 \times 0.3 = 0.6$
$2 \times 1.3 = 2 + 0.6$
$= 2.6$

Method 2

$2 \times 1.3 = 2 \times 13$ tenths
$= 26$ tenths
$= 2.6$

Charles' dad should use 2.6 l of vegetable stock.

Guided Practice

1 Multiply.

(a) $3 \times 0.3 =$

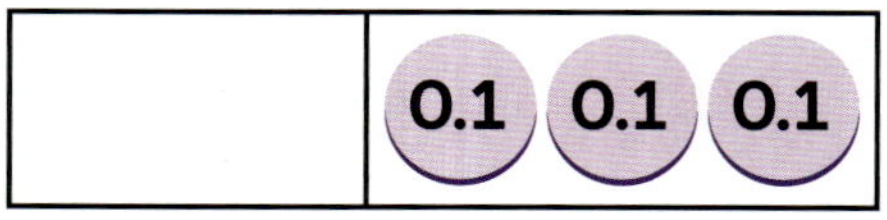

(b) $3 \times 0.12 =$

(c) $4 \times 2.1 =$

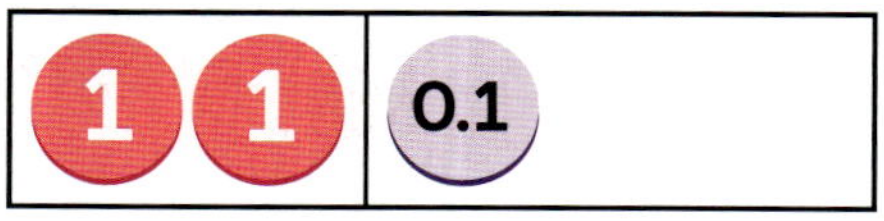

2 Multiply.

(a) $5 \times 0.1 =$

(b) $3 \times 0.21 =$

(c) $2 \times 2.3 =$

3

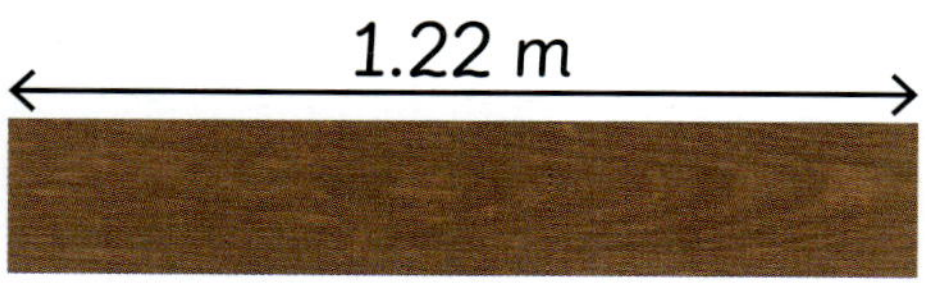

Ruby has 3 pieces of wood. Each piece of wood has the same length of 1.22 m.
What is the total length of the wood if the 3 pieces are arranged end-to-end?

Complete Worksheet **6**: pages **113–114**

Multiplying Decimals with Renaming

Lesson 7

Explore

Mr Nightingale walked 1.32 km every day for 6 days. What is the total distance Mr Nightingale walked over the 6 days?

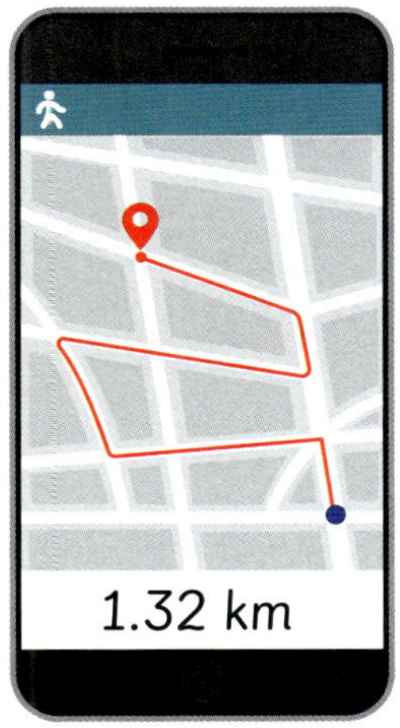

Master

1 6 × 1.32 km =

6 × 1.32 = 1.32 × 6

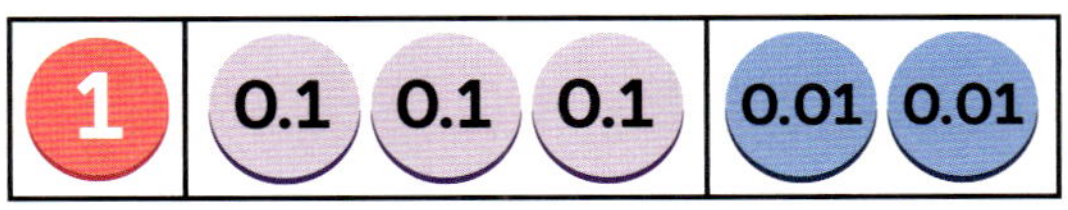

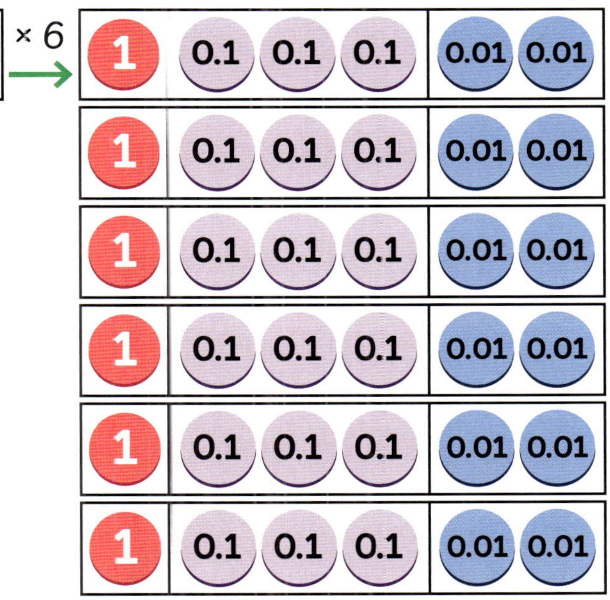

$$
\begin{aligned}
6 \times 1 &= 6 \\
6 \times 0.3 &= 1.8 \\
6 \times 0.02 &= 0.12 \\
6 \times 1.32 &= 6 + 1.8 + 0.12 \\
&= 7.92
\end{aligned}
$$

6 × 3 tenths = 18 tenths

18 tenths = 1.8

6 × 2 hundredths = 12 hundredths

12 hundredths = 0.12

2 6×1.32 km = ______

$6 \times 1.32 = 6 \times 132$ hundredths
$= 792$ hundredths
$= 7.92$

$$\begin{array}{rrrr} & {}^{1}1 & {}^{1}3 & 2 \\ \times & & & 6 \\ \hline & 7 & 9 & 2 \\ \hline \end{array}$$

Mr Nightingale walked 7.92 km over the 6 days.

Guided Practice

1 Multiply.

(a) $3 \times 2.5 =$ ______

(b) $5 \times 1.3 =$ ______

(c) $6 \times 0.12 =$ ______

(d) $2.4 \times 4 =$ ______

(e) $7 \times 1.31 =$ ______

(f) $2.12 \times 7 =$ ______

2 Ravi's mum used 4 tins of paint to paint the garden fence.
Each tin holds 2.79 l of paint.
What is the total volume of paint that she used?

Complete Worksheet **7**: pages **115–116**

Multiplying Decimals with Renaming

Lesson 8

Explore

Holly's school built 6 new classrooms.
In each classroom, an electrician used 7.23 m of wire for the lights.
What was the total length of wire used in the 6 classrooms?

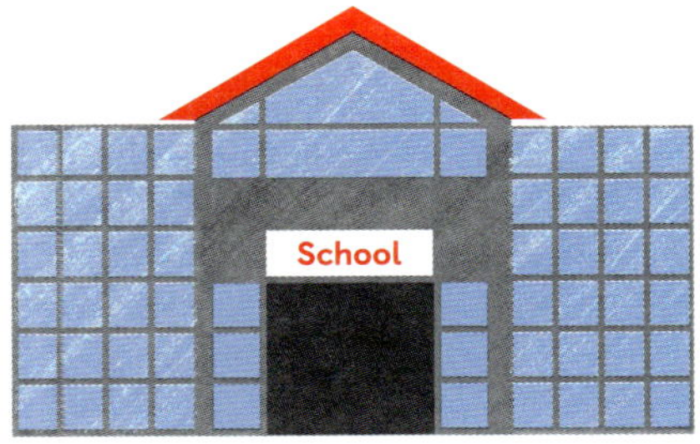

Master

6 × 7.23 m =

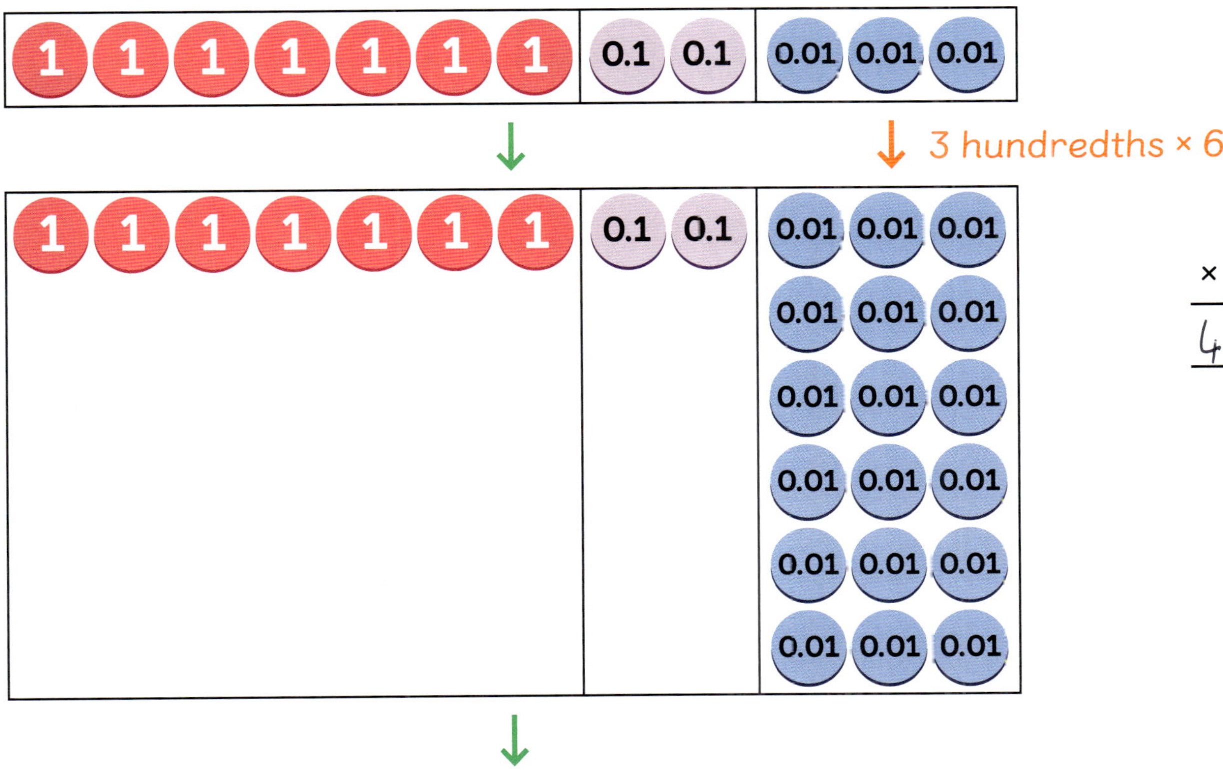

	1		1	
	7	.	2	3
×				6
4	3	.	3	8

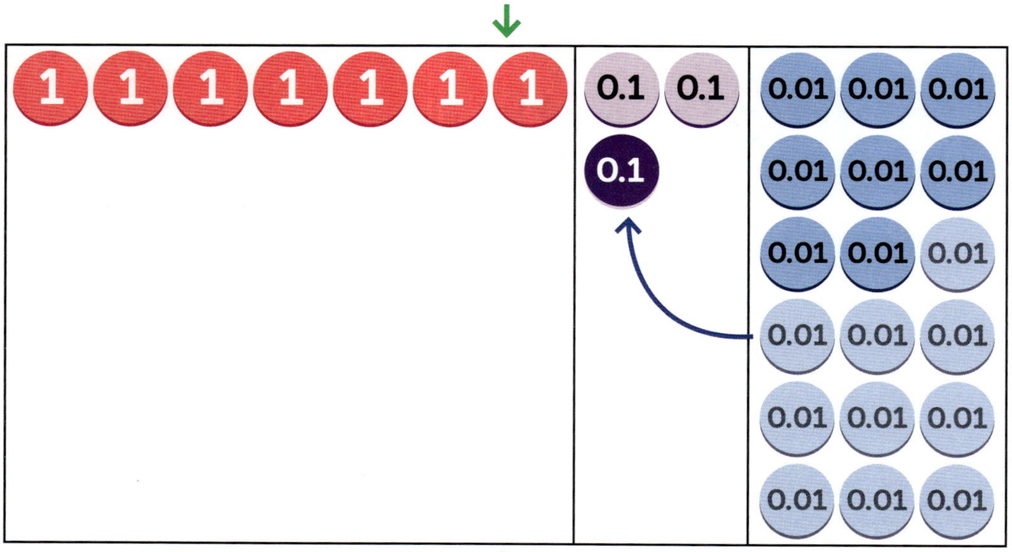

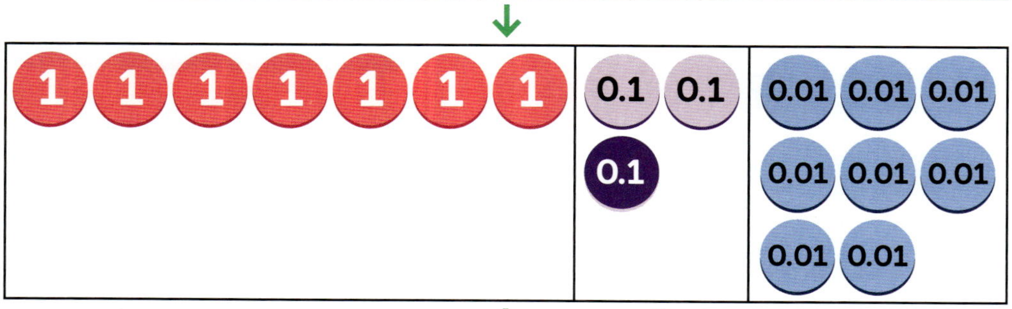

$$\begin{array}{r} 7.\overset{1}{2}\,3 \\ \times \quad 6 \\ \hline 8 \\ \hline \end{array}$$

2 tenths × 6 + 1 tenth

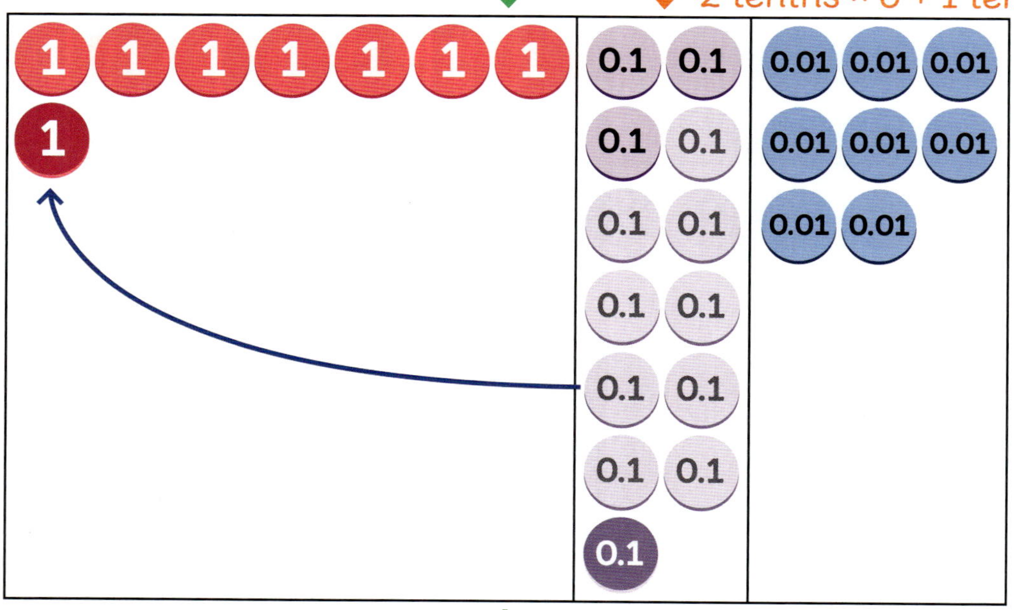

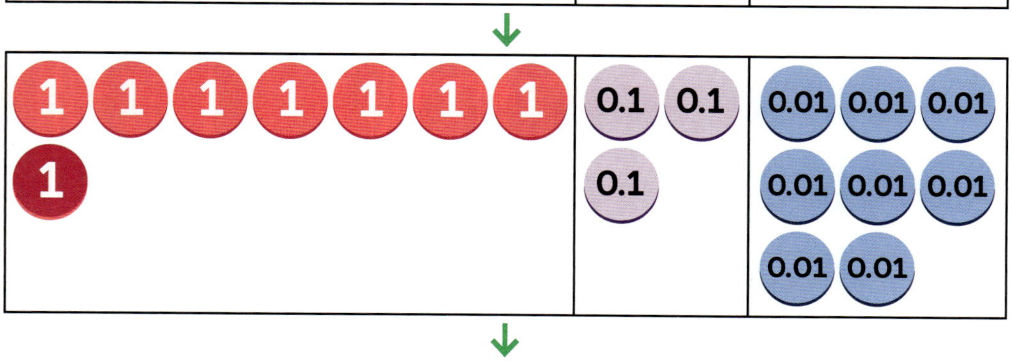

$$\begin{array}{r} \overset{1}{7}.\overset{1}{2}\,3 \\ \times \quad 6 \\ \hline 3\,8 \\ \hline \end{array}$$

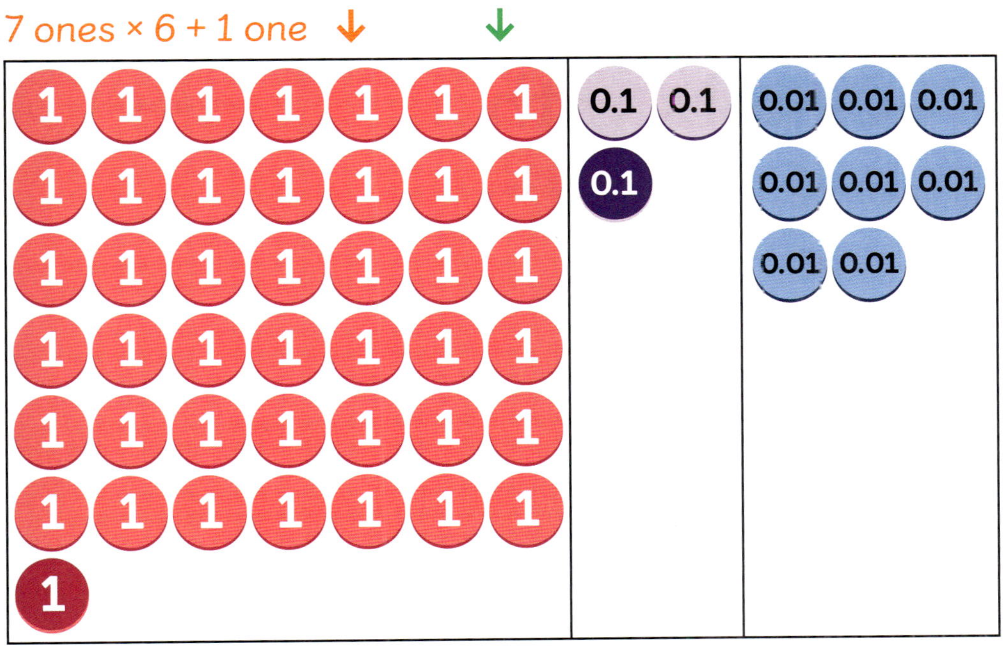

$$\begin{array}{r} {}^{1}7\,.\,{}^{1}2\;3 \\ \times \quad\quad 6 \\ \hline 4\;3\,.\,3\;8 \\ \hline \end{array}$$

6 × 7.23 = 43.38

The total length of wire used in the 6 classrooms was 43.38 m.

Guided Practice

Multiply.

1

$$\begin{array}{r} 2\,.\,1\;3 \\ \times \quad\quad 4 \\ \hline \\ \hline \end{array}$$

2

$$\begin{array}{r} 7\,.\,1\;5 \\ \times \quad\quad 8 \\ \hline \\ \hline \end{array}$$

3

$$\begin{array}{r} 1\;2\,.\,6\;4 \\ \times \quad\quad 7 \\ \hline \\ \hline \end{array}$$

4

$$\begin{array}{r} 3\;7\,.\,8\;9 \\ \times \quad\quad 5 \\ \hline \\ \hline \end{array}$$

Complete Worksheet **8**: pages **117–118**

Multiplying Decimals with Renaming

Lesson 9

Explore

Lulu wants to make a vegetable traybake.

Vegetable traybake (Serves 4 people)

0.15 kg	peppers	1 tsp	salt
0.51 kg	onions	1.05 kg	tomatoes
1 tbsp	garlic	1.54 kg	potatoes

She needs to make the recipe 3 times greater so that it feeds 12 people. What are the measurements she needs to use?

Master

1 3×0.15 kg = ▭

$$
\begin{array}{r}
0.15 \\
\times \quad 3 \\
\hline
0.15 \\
+\ 0.30 \\
\hline
0.45 \\
\hline
\end{array}
$$

0.15 → 3×0.05

0.30 → 3×0.1

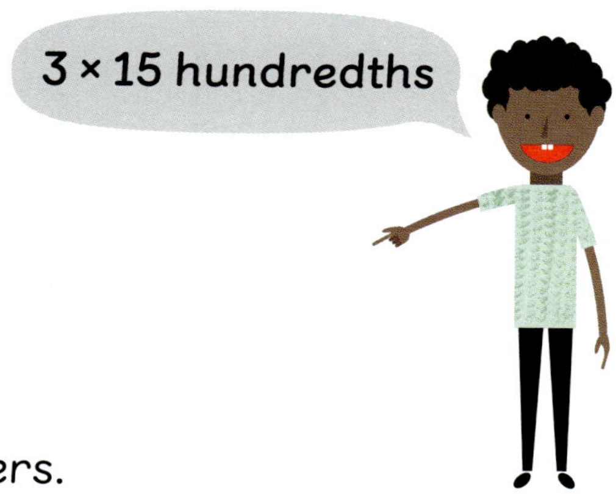

Lulu needs to use 0.45 kg of peppers.

2 3×1.05 kg = ▭

$$\begin{array}{rrrr} & 1 & . & 0\ 5 \\ \times & & & 3 \\ \hline & 0 & . & 1\ 5 \\ + & 3 & . & 0\ 0 \\ \hline & 3 & . & 1\ 5 \\ \hline \end{array}$$

→ 3×0.05

→ 3×1

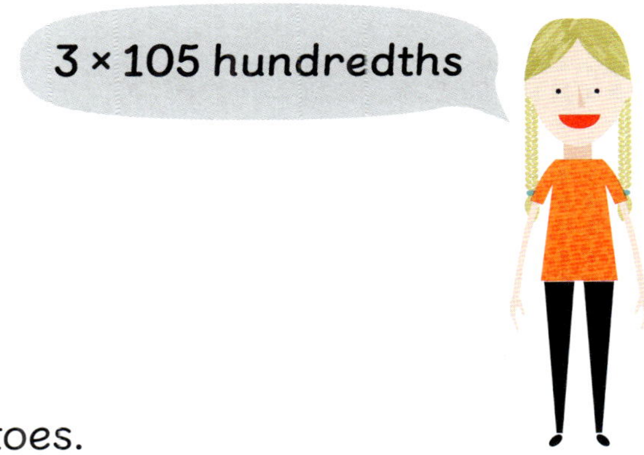

Lulu needs to use 3.15 kg of tomatoes.

3 3×0.51 kg = ▭

$$\begin{array}{rrrr} & 0 & . & 5\ 1 \\ \times & & & 3 \\ \hline & 0 & . & 0\ 3 \\ + & 1 & . & 5\ 0 \\ \hline & 1 & . & 5\ 3 \\ \hline \end{array}$$

$$\begin{aligned} 3 \times 0.51 &= 0.03 + 1.5 \\ &= 1.53 \end{aligned}$$

Lulu needs to use 1.53 kg of onions.

4 3×1.54 kg = ▭

$$\begin{array}{rrrr} & \overset{1}{1} & \overset{1}{.} & 5\ 4 \\ \times & & & 3 \\ \hline & 4 & . & 6\ 2 \\ \hline \end{array}$$

$$\begin{aligned} 3 \times 0.04 &= 0.12 \\ 3 \times 0.5 &= 1.5 \\ 3 \times 1 &= 3 \\ \hline 3 \times 1.54 &= 4.62 \end{aligned}$$

Lulu needs to use 4.62 kg of potatoes.

Activity Time

Partner work

You will need:

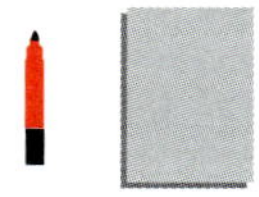

(1) 1.23 × 9 = ____

Can you use 1.23 × 10 to help you calculate 1.23 × 9?

(2) 1.23 × 5 = ____

Can you use 1.23 × 10 to help you calculate 1.23 × 5?

(3) 1.23 × 4 = ____

What methods can you use to solve this equation?
Can you use your answers from (1) and (2)?

Guided Practice

1 Multiply.

(a) 4 × 1.22 = ____

```
   1 . 2 2
×        4
----------
   ____      → 4 × 0.02
   ____      → 4 × ____
   ____      → 4 × ____
----------
   ____
----------
```

(b) $6 \times 2.43 =$ ____

$$\begin{array}{r} 2\,.\,4\,3 \\ \times \quad\quad 6 \\ \hline \square \\ \square \\ \square \\ \hline \square \\ \hline \end{array}$$

→ 6 × ____

→ 6 × ____

→ 6 × ____

2 Multiply.

(a) $3.18 \times 3 =$ ____

(b) $4.74 \times 8 =$ ____

3 A restaurant used 8 bags of onions in one evening.
Each bag of onions had a mass of 2.89 kg.
What is the total mass of onions the restaurant used?

Complete Worksheet **9**: pages **119–120**

Dividing Decimals without Renaming

Lesson 10

Explore

8.42 kg of compost is divided equally between 2 containers.
What is the mass of the compost in each container?

Master

1 8.42 kg ÷ 2 = ______

8.42

4.21

4.21

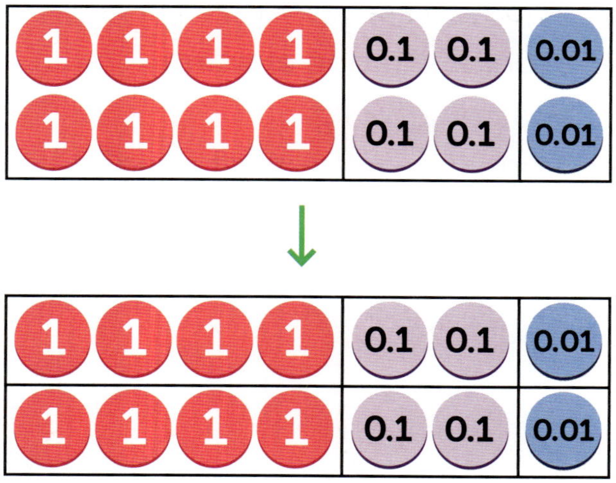

8 ones ÷ 2 = 4 ones
4 tenths ÷ 2 = 2 tenths
2 hundredths ÷ 2 = 1 hundredth

$8 \div 2 = 4$
$0.4 \div 2 = 0.2$
$0.02 \div 2 = 0.01$

$8.42 \div 2 = 4.21$

2 8.42 kg ÷ 2 = [ ]

```
       [ ] . [ ] [ ]
  2 )   8  .  4   2
      - 8              →  2 × 4
      ----------------
        0  .  4
      - 0  .  4        →  2 × 0.2
      ----------------
        0  .  0   2
      - 0  .  0   2    →  2 × 0.01
      ----------------
                  0
```

8.42 ÷ 2 = 4.21

There are 4.21 kg of compost in each container.

Guided Practice

Divide.

1 6.4 ÷ 2 = [ ]

2 9.6 ÷ 3 = [ ]

3 5.05 ÷ 5 = [ ]

4 2.86 ÷ 2 = [ ]

5 12.8 ÷ 4 = [ ]

6 63.93 ÷ 3 = [ ]

Complete Worksheet **10**: pages **121–122**

Dividing Decimals with Renaming

Lesson 11

Explore

Amira has four times as much juice as Ravi.
Together they have 6.15 l of juice.
How much juice does Ravi have?

Master

1 6.15 l ÷ 5 = ▭

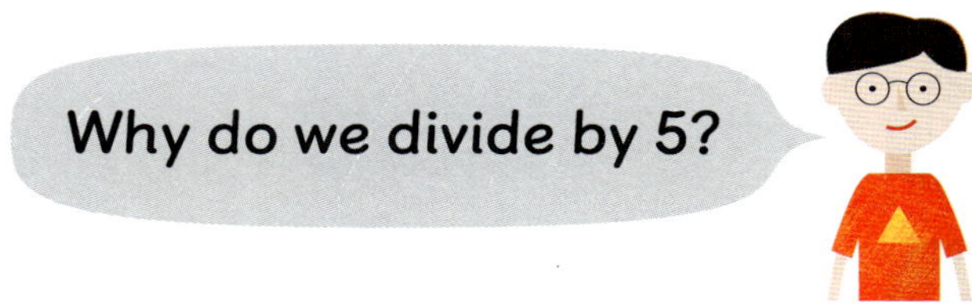

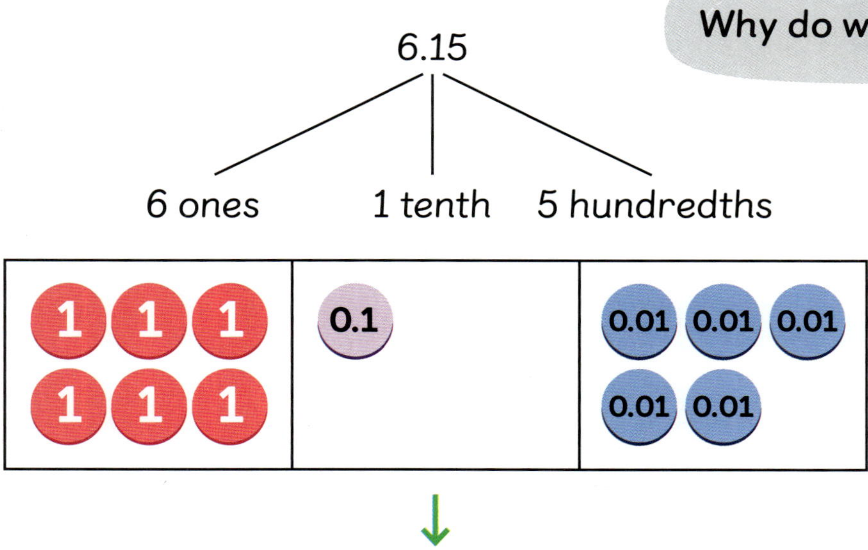

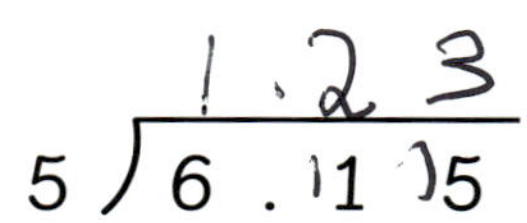

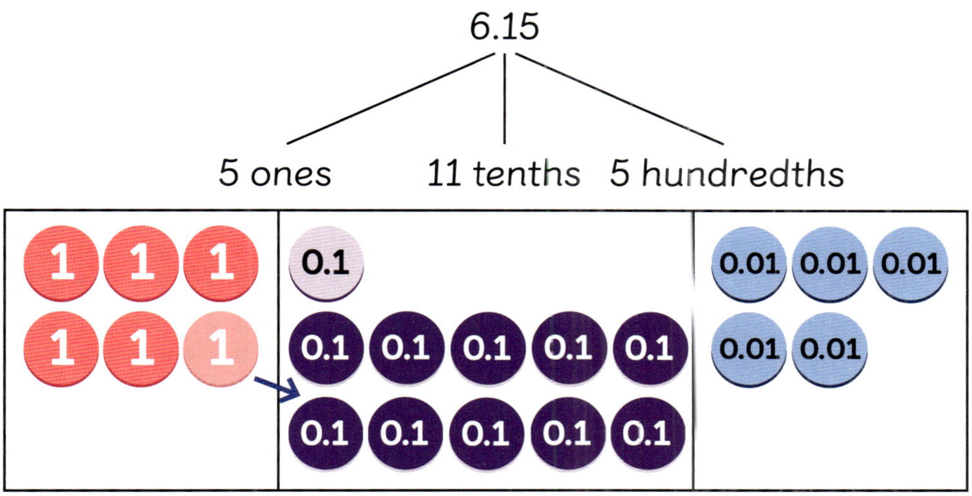


```
      1
5 ) 6 . 1 5
  − 5
    1 . 1 5
```

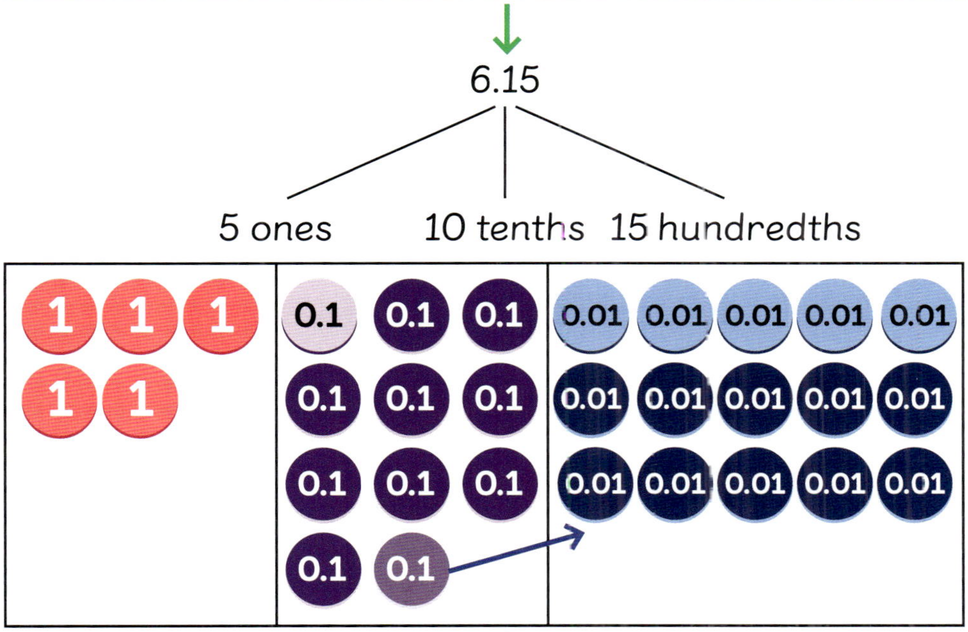

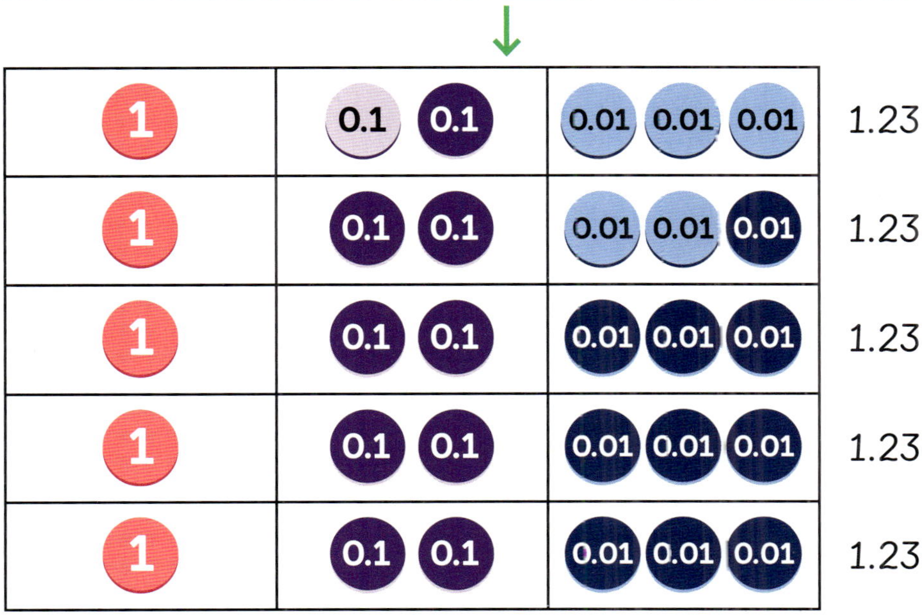


```
      1 . 2 3
5 ) 6 . 1 5
  − 5
    1 . 1 5
  − 1 . 0
    0 . 1 5
  − 0 . 1 5
          0
```

$6.15 \div 5 = 1.23$

2

6.15

6 ones	1 tenth	5 hundredths
↓	↓	↓
5 ones	11 tenths	5 hundredths
↓	↓	↓
5 ones	10 tenths	15 hundredths
↓ ÷ 5	↓ ÷ 5	↓ ÷ 5
1 one	2 tenths	3 hundredths

$6.15 \div 5 = 1.23$

3

$$
\begin{array}{r|l}
 & 1.23 \\
5 \,) & 6.15 \\
 & -5 \qquad \longrightarrow 5 \times 1 \\
\hline
 & 1.15 \\
 & -1.0 \qquad \longrightarrow 5 \times 0.2 \\
\hline
 & 0.15 \\
 & -0.15 \qquad \longrightarrow 5 \times 0.03 \\
\hline
 & 0
\end{array}
$$

$6.15 \div 5 = 1.23$

Ravi has 1.23 l of juice.

4 Amira combined the juice in a punch bowl, then poured all of it equally into 4 cups.
Find the volume of juice in each cup.

```
      1 . 5 3 7 5
3 )   6 . 1 5 0 0
    - 4                  →  4 × 1
      2 . 1
    - 2 . 0              →  4 × 0.5
      0 . 1 5
    - 0 . 1 2            →  4 × 0.03
      0 . 0 3 0
    - 0 . 0 2 8          →  4 × 0.007
      0 . 0 0 2 0
    - 0 . 0 0 2 0        →  4 × 0.0005
                0
```

1.5375 is 1.5 when rounded to 1 decimal place.

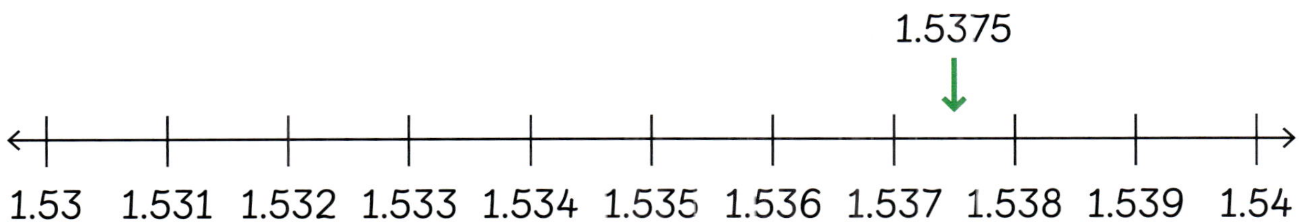

1.5375 is closer to 1.54 than 1.53.
1.5375 is 1.54 when rounded to 2 decimal places.

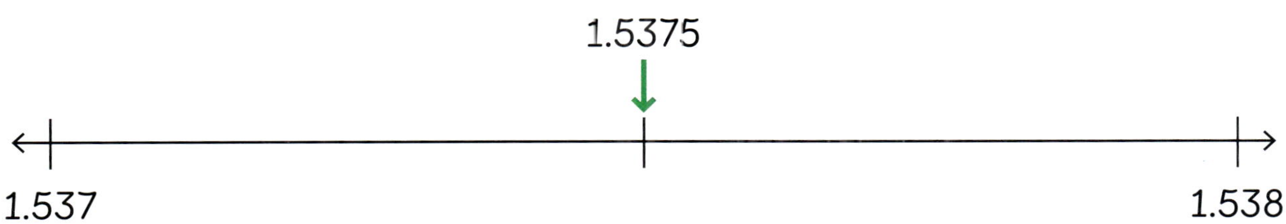

1.5375 is exactly halfway between 1.537 and 1.538.
Round up to the greater decimal.
1.5375 is 1.538 when rounded to 3 decimal places.

Guided Practice

1 $9.42 \div 3 = \square$

$$
\begin{array}{r@{}l}
 & \square.\square\square \\
3\,\overline{)\,9.42} & \\
-\,9 & \quad \rightarrow 3 \times \square \\
\hline
0.42 & \\
-\,0.3 & \quad \rightarrow 3 \times \square \\
\hline
0.12 & \\
-\,0.12 & \quad \rightarrow 3 \times \square \\
\hline
0 &
\end{array}
$$

2 Divide. Give your answer correct to 2 decimal places.

(a) $6.87 \div 4 = \square$

(b) $9.96 \div 7 = \square$

3 Divide. Give your answer correct to 3 decimal places.

(a) $4.63 \div 3 = \square$

(b) $7.09 \div 8 = \square$

4 A painter used 7.84 l of paint equally on 6 identical flower boxes. What is the volume of paint used for each flower box? Give your answer correct to 2 decimal places.

Complete Worksheet **11**: pages **123–124**

Multiplying a Decimal by a 2-Digit Whole Number

Lesson 12

Explore

Ruby wants to put a fence around a vegetable plot.
She has 28 fence panels.
Each panel is 1.2 m wide.
What is the total length of fencing she will use?

Master

1 1.2 m × 28 = ______

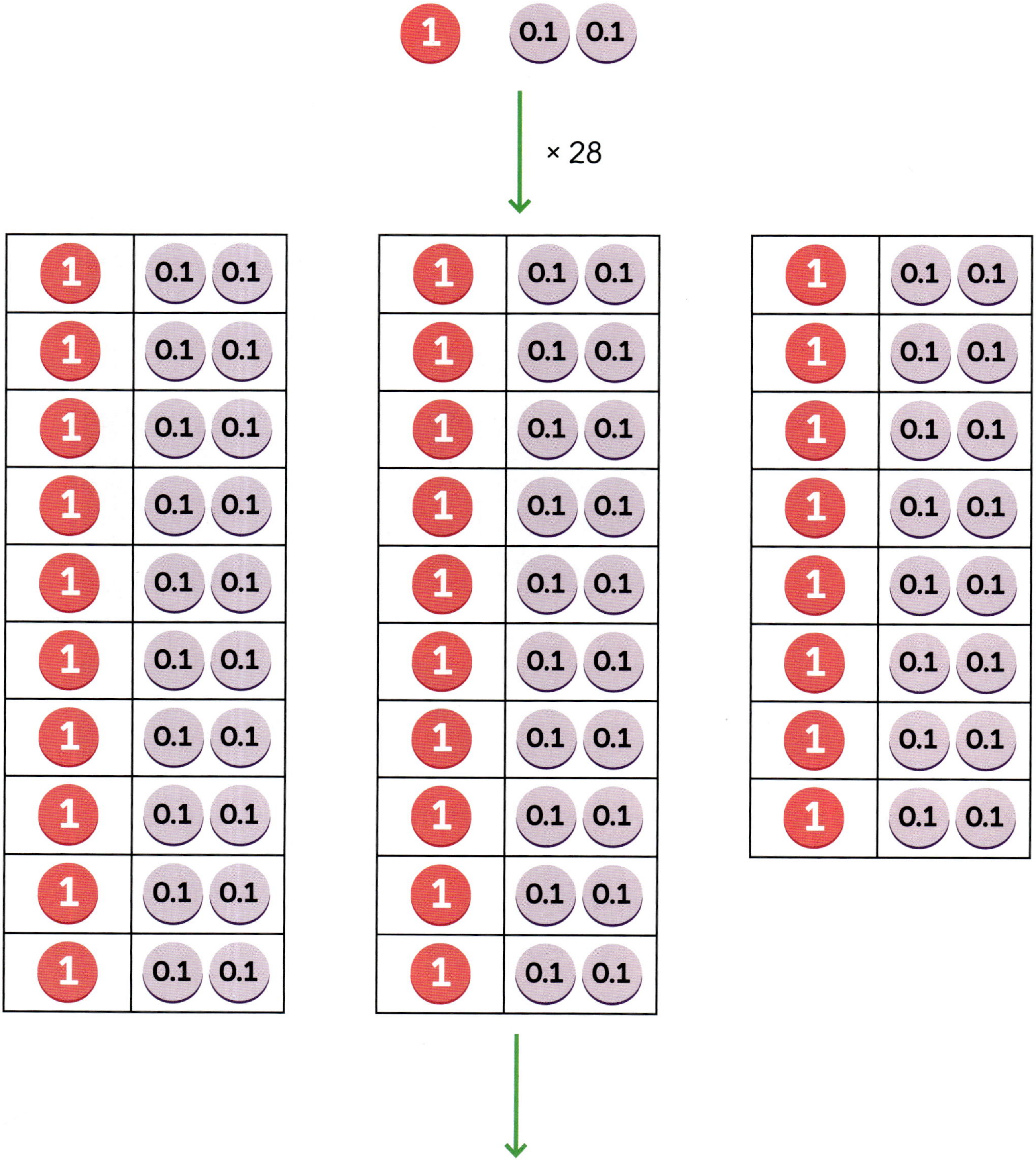

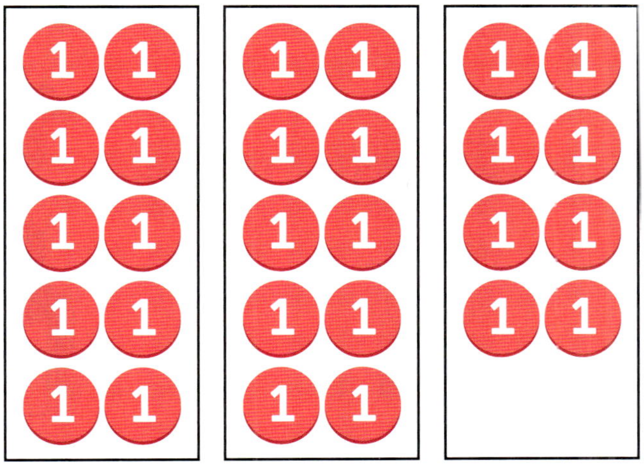

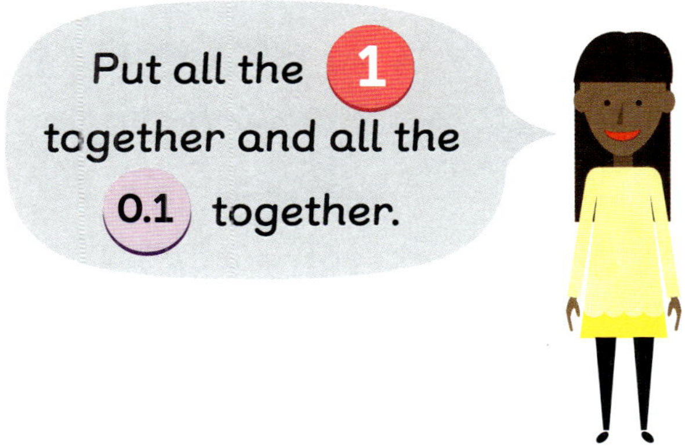

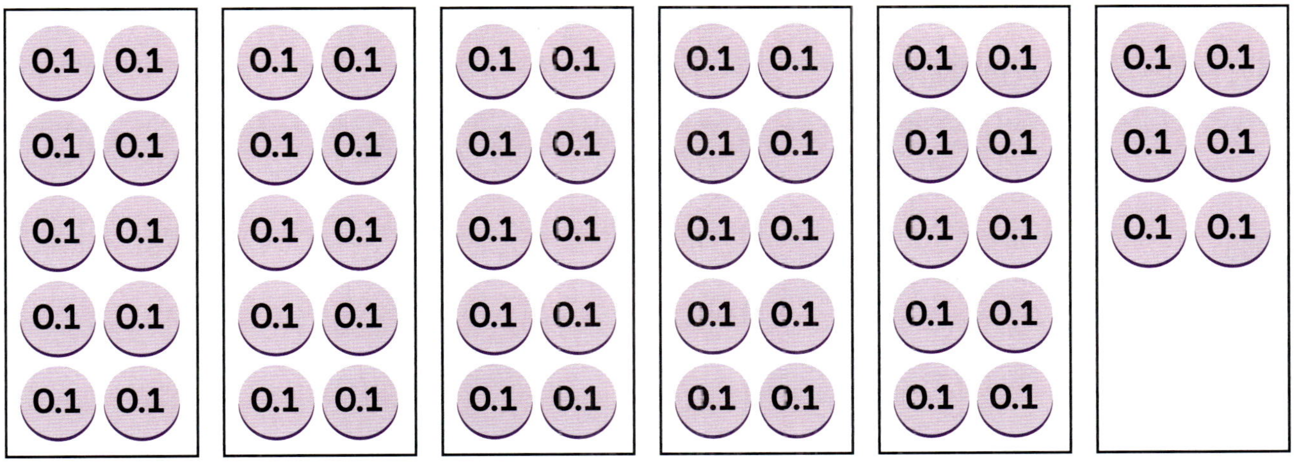

$1 \times 28 = 28$

$0.2 \times 28 = 5.6$

$1.2 \times 28 = 28 + 5.6$

$= 33.6$

2 $1.2 \times 28 = (1.2 \times 20) + (1.2 \times 8)$

$1.2 \times 10 = 12$

$1.2 \times 20 = 24$

$1.2 \times 8 = 9.6$

$1.2 \times 28 = 24 + 9.6$

$= 33.6$

	1	.	2
×			8
	1	.	6
+	8	.	0
	9	.	6

3 1.2 m × 28 = ____

1.2 × 14 = 1 × 14 + 0.2 × 14
= 14 + 2.8
= 16.8

1.2 × 28 = 16.8 + 16.8 = 33.6

The total length of fencing she will use is 33.6 m.

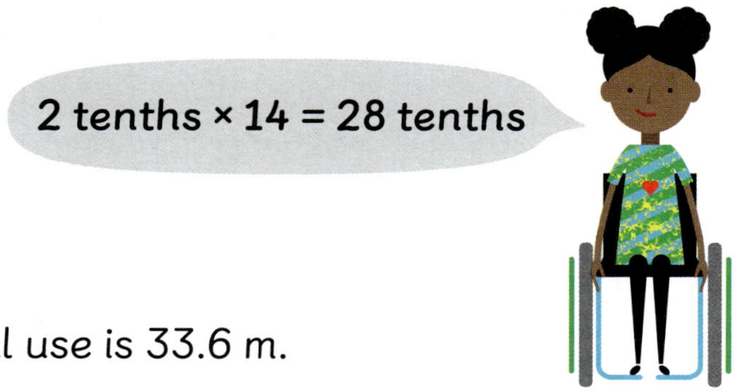

Guided Practice

Multiply.

1 (a) 3.2 × 10 = ____

(b) 3.2 × 20 = ____

(c) 1.6 × 20 = ____

(d) 1.6 × 40 = ____

2 (a) 4.5 × 12 = ____

(b) 4.5 × 32 = ____

(c) 9.01 × 12 = ____

(d) 9.01 × 25 = ____

Complete Worksheet **12**: pages **125–126**

Dividing a Decimal by a 2-Digit Whole Number

Lesson 13

Explore

Elliott's school has organised a cross-country relay.

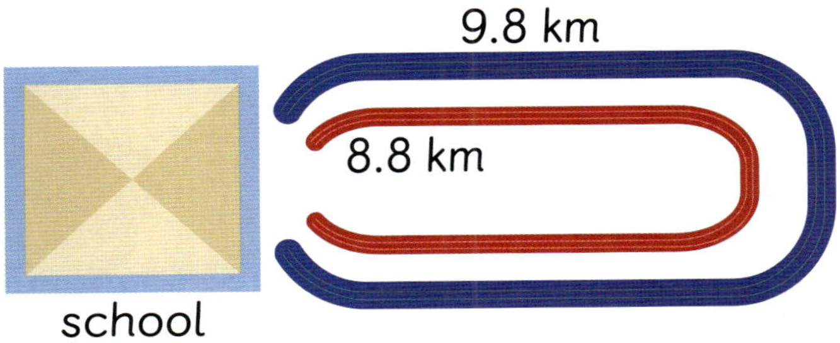

The blue track has 14 runners per team.
The red track has 11 runners per team.
What distance will each teammate run on each track?

Master

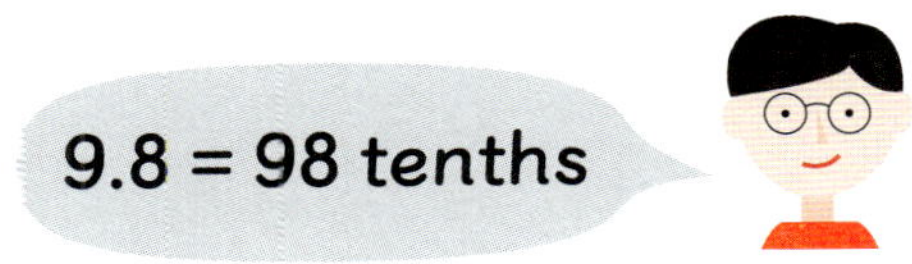

1 $9.8 \text{ km} \div 14 =$

$9.8 \div 14 = 98 \text{ tenths} \div 14$
$= 7 \text{ tenths}$
$= 0.7$

On the blue track, each teammate will run 0.7 km.

2 $8.8 \text{ km} \div 11 =$

$8.8 \div 11 = 88 \text{ tenths} \div 11$
$= 8 \text{ tenths}$
$= 0.8$

On the red track, each teammate will run 0.8 km.

Guided Practice

1 Divide.

(a) $0.36 \div 12 =$

(b) $0.72 \div 12 =$

(c) $0.75 \div 15 =$

2 Find the value of:

(a) $2.73 \div 13 =$

(b) $2.94 \div 14 =$

(c) $1.95 \div 15 =$

3 A painter used 32.48 l of paint to paint 16 classrooms. She used the same volume of paint in each classroom. What was the volume of paint used in a single classroom?

Complete Worksheet **13**: pages **127–128**

Dividing a Decimal by a 2-Digit Whole Number

Lesson 14

Explore

A restaurant has 4.65 kg of pizza dough. The chef uses all of the dough to make 15 pizzas. Each pizza uses the same amount of dough.

What is the mass of the dough used in a single pizza?

Master

1 4.65 kg ÷ 15 = ____

4.65

4.5 = 45 tenths → ÷ 15 → = 3 tenths

0.15 = 15 hundredths → ÷ 15 → = 1 hundredth

3 tenths + 1 hundredth = 0.3 + 0.01
= 0.31

4.65 ÷ 15 = 0.31

2 4.65 kg ÷ 15 = ____

```
      _ . _ _
15 ) 4 . 6 5
   - 4 . 5      → 15 × ____
     0 . 1 5
   - 0 . 1 5    → 15 × ____
           0
```

4.65

4.5 0.15

A single pizza uses 0.31 kg of the dough.

Guided Practice

1 Which method below would you use to help you divide 9.6 by 12?

(a) $12\overline{)9.6}$

(b)

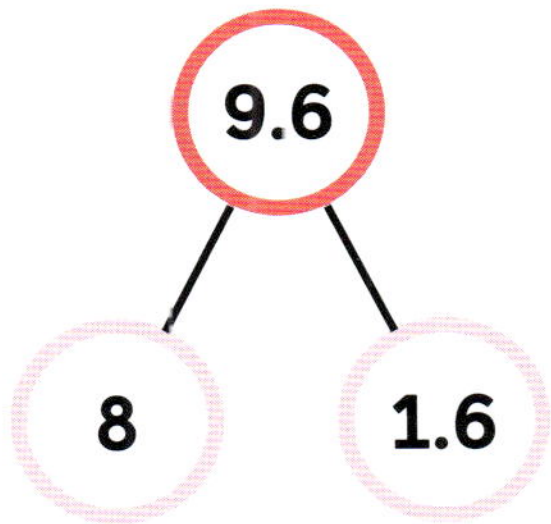

(c)

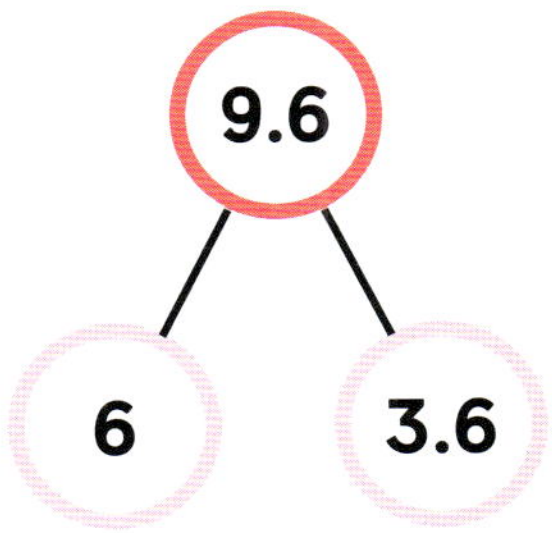

(d)

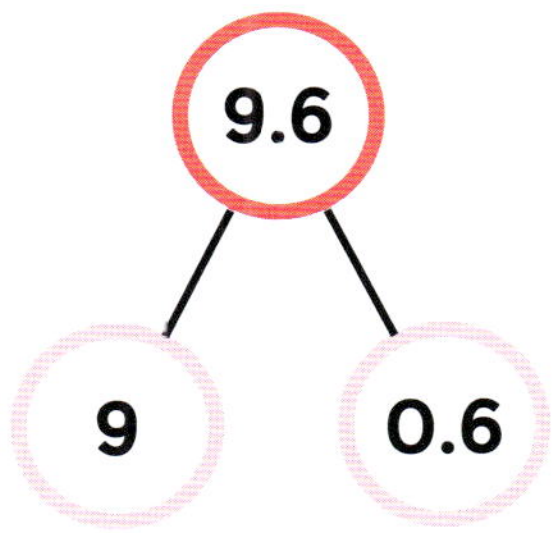

2 (a) 1.44 ÷ 12 = ____

(b) Use part (a) to find 2.88 ÷ 12.

(c) Use part (b) to find 2.88 ÷ 6.

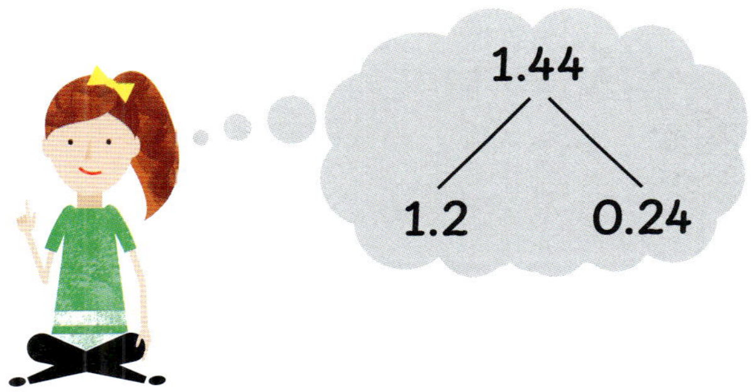

3 Divide.

(a) 4.48 ÷ 14 = ____

$$\begin{array}{r} \square.\square\square \\ 14\overline{)4.48} \end{array}$$

(b) 6.72 ÷ 16 = ____

$$\begin{array}{r} \square.\square\square \\ 16\overline{)6.72} \end{array}$$

Complete Worksheet **14**: pages **129–130**

Mind Challenge

Arrange the digits 2, 4, 6 and 0 to make a product.

$\square . \square \square \times \square$

How many different products are there?
What is the greatest product?
What is the smallest product?

My Maths Journal

Find the values of 2.48 × 12 and 248 × 12. Can you see how knowing one answer can help you find the other answer?

If you know that 2.64 ÷ 6 = 0.44, how can you use this to find the value of 26.4 ÷ 6?

Review

I can...

- ☐ read and write decimals.
- ☐ identify the value of digits in decimal places.
- ☐ relate division of whole numbers to fractions and decimals.
- ☐ convert fractions to decimals.
- ☐ convert decimals to fractions.
- ☐ multiply and divide decimals by 1-digit and 2-digit numbers.
- ☐ round a decimal number to 1, 2 or 3 decimal places.

What measurements do we take at home?

Chapter 5
Measurements

Converting Units of Length: Millimetres and Centimetres

Lesson 1

Explore

Hannah measures her picture.
Will it fit into her frame?

Master

1 Hannah writes 20 cm and 12.5 cm in millimetres.

1 cm = 10 mm
20 cm = 20 × 10 mm
= 200 mm
12.5 cm = 12.5 × 10 mm
= 125 mm

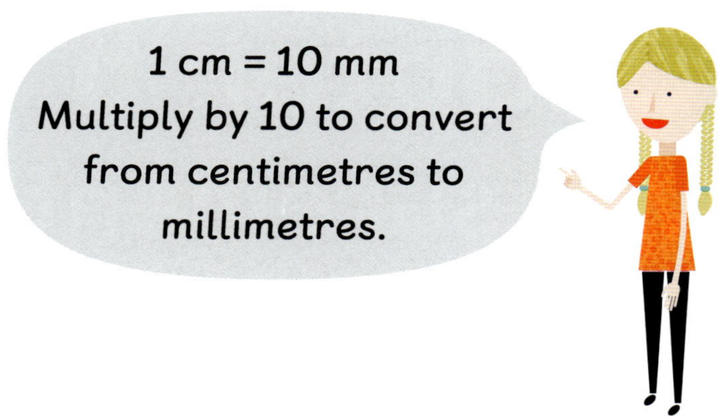

Hannah's picture measures
125 mm by 200 mm.

The frame measures
130 mm by 210 mm.

130 mm is greater than 125 mm, and
210 mm is greater than 200 mm.

2 Write 130 mm and 210 mm in centimetres.

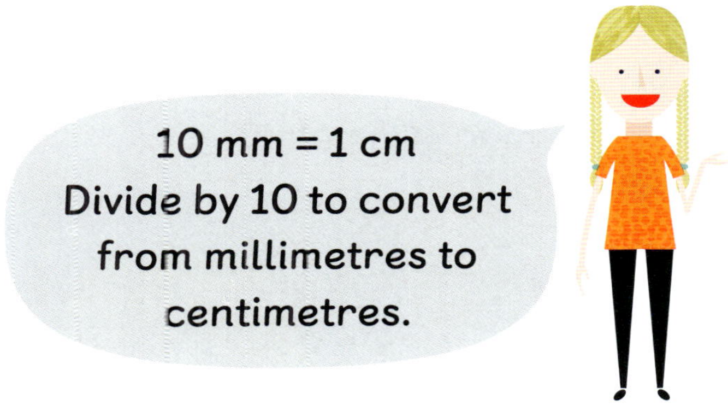

10 mm = 1 cm
130 mm = 130 ÷ 10 mm
= 13 cm

210 mm = 210 ÷ 10 mm
= 21 cm

13 cm is greater than 12.5 cm, and 21 cm is greater than 20 cm.

Hannah's picture will fit in the frame.

3 Express the width of Hannah's picture in centimetres and millimetres.

12.5 cm = ___ cm ___ mm

0.5 cm = 0.5 × 10 mm

= 5 mm

12.5 cm = 12 cm 5 mm

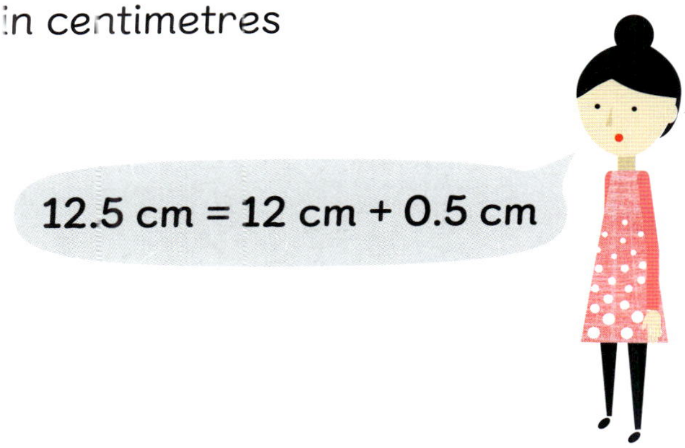

Guided Practice

1 Write in millimetres.

(a) 2 cm = ☐ mm (b) 3.4 cm = ☐ mm

2 Write in centimetres.

(a) 6 mm = ☐ cm (b) 408 mm = ☐ cm

3 Write in centimetres and millimetres.

(a) 10.1 cm = ☐ cm ☐ mm (b) 515 mm = ☐ cm ☐ mm

4 Measure the sides of these drawings in centimetres and convert them to millimetres.

5 Calculate the perimeter of each drawing in millimetres and convert to centimetres.
How can you check to see if your answers are correct?

Complete Worksheet **1**: pages **135–136**

Converting Units of Length: Metres and Centimetres

Lesson 2

Explore

Sam needs different lengths of fabric to make a costume for a school play.

Blue fabric: 1.25 m

White fabric: 2 m 45 cm

Red fabric: 95 cm

What is the total length of fabric he needs?

Master

1 Convert 1.25 m to centimetres.

$$1 \text{ m} = 100 \text{ cm}$$
$$1.25 \text{ m} = 1.25 \times 100 \text{ cm}$$
$$= 125 \text{ cm}$$

2 Convert 2 m 45 cm to centimetres.

$$2 \text{ m} = 200 \text{ cm}$$
$$2 \text{ m } 45 \text{ cm} = 200 + 45 \text{ cm}$$
$$= 245 \text{ cm}$$

3. Find the total length in centimetres.

125 cm + 245 cm + 95 cm = 465 cm
Sam needs 465 cm of fabric in total.

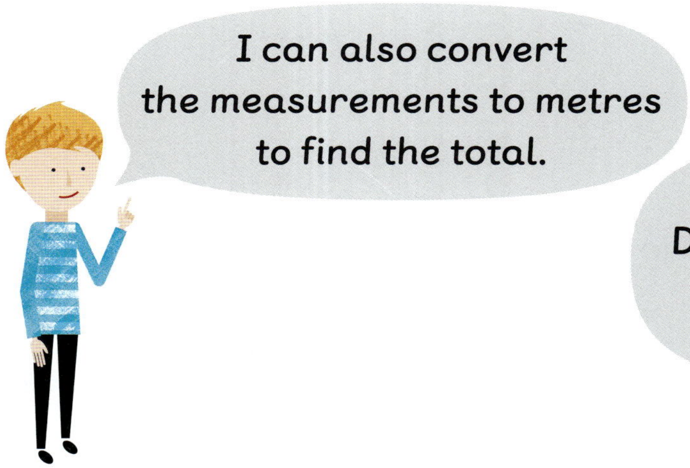

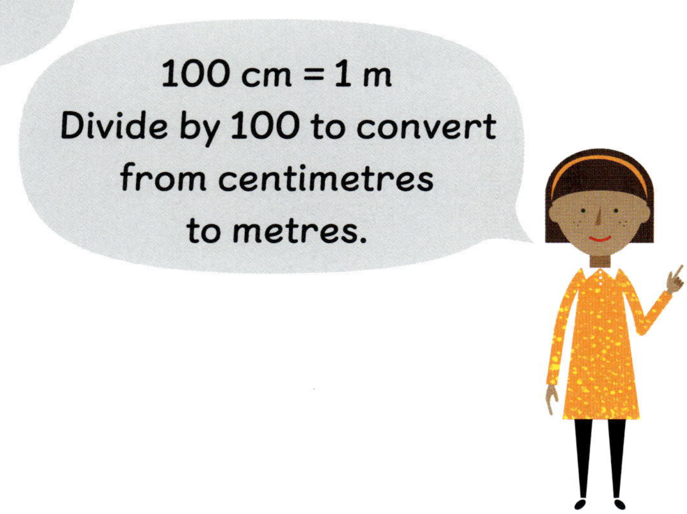

4. Convert 95 centimetres to metres.

100 cm = 1 m
95 cm = 95 ÷ 100 m
= 0.95 m

5. Convert 2 m 45 cm to metres.

100 cm = 1 m
45 cm = 45 ÷ 100 m
= 0.45 m
2 m 45 cm = 2 m + 0.45 m
= 2.45 m

6. Find the total length in metres.

0.95 m + 2.45 m + 1.25 m = 4.65 m
Sam needs 4.65 m of fabric in total.

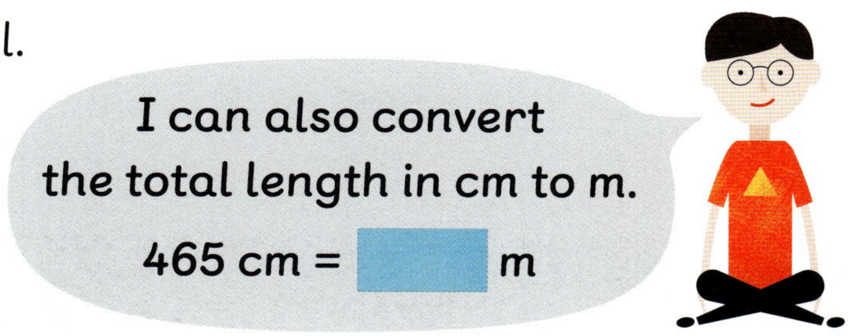

Guided Practice

1 Write in centimetres.

(a) 3 m = ☐ cm

(b) 2.7 m = ☐ cm

(c) 4.45 m = ☐ cm

2 Write in metres.

(a) 500 cm = ☐ m

(b) 105 cm = ☐ m

(c) 22 cm = ☐ m

3 Fill in the blanks.

(a) 1 m 18 cm = ☐ cm

(b) 3 m ☐ cm = 306 cm

(c) ☐ m ☐ cm = 1010 cm

4 Fill in the blanks.

(a) 4 m ☐ cm = 4.15 m

(b) ☐ m ☐ cm = 7.8 m

(c) 6 m 9 cm = ☐ m

Complete Worksheet **2**: pages **137–138**

Converting Units of Length: Kilometres and Metres

Lesson 3

Explore

Holly is practising for the school sports day.
She runs each day for three days.
How many metres does Holly run in total?

Holly's running distances	
Monday	1.8 km
Tuesday	2 km 25 m
Wednesday	3.05 km

Master

1 Express 1.8 km in metres.

1 km = 1000 m
1.8 km = 1.8 × 1000 m
= 1800 m

2 Express 2 km 25 m in metres.

2 km = 2000 m
2 km 25 m = 2000 m + 25 m
= 2025 m

3 Express 3.05 km in metres.

1 km = 1000 m
3.05 km = 3.05 × 1000 m
= 3050 m

4 Find the total distance in metres.

1800 m + 2025 m + 3050 m = 6875 m
Holly runs 6875 m in total.

5 Express the distance Holly runs on Tuesday in kilometres.

2 km 25 m = ____ km

1000 m = 1 km
25 m = 25 ÷ 1000 km
= 0.025 km
2 km 25 m = 2 km + 0.025 km
= 2.025 km

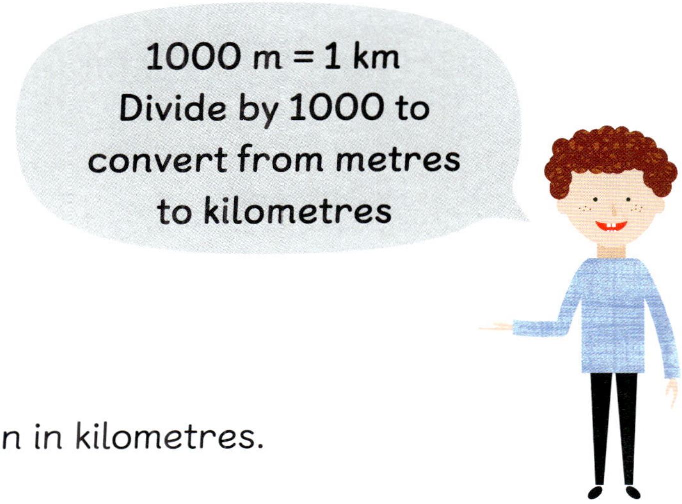

6 Express the total distance Holly ran in kilometres.

1000 m = 1 km
6875 m = 6875 ÷ 1000 km
= 6.875 km

7 Express the total distance Holly ran in kilometres and metres.

6.875 km = ____ km ____ m

0.875 km = 0.875 × 1000 m
= 875 m
6.875 km = 6 km 875 m

Guided Practice

1 Write in metres.

(a) 10 km = ____ m

(b) 3.8 km = ____ m

(c) 4.05 km = ____ m

2 Write in kilometres.

(a) 80 m = ____ km

(b) 5009 m = ____ km

(c) 2400 m = ____ km

3 Fill in the blanks.

(a) 2 km 632 m = ____ m

(b) 7 km ____ m = 7077 m

(c) ____ km ____ m = 14 100 m

4 Fill in the blanks.

(a) 6 km ____ m = 6.807 km

(b) ____ km ____ m = 9.03 km

(c) 10 km 1 m = ____ km

Complete Worksheet **3**: pages **139–140**

Converting Units of Length: Miles and Kilometres

Lesson 4

Explore

Sheffield to Derby is the shortest distance.

Sheffield to Derby
49 miles

Swansea to Cardiff
69 km

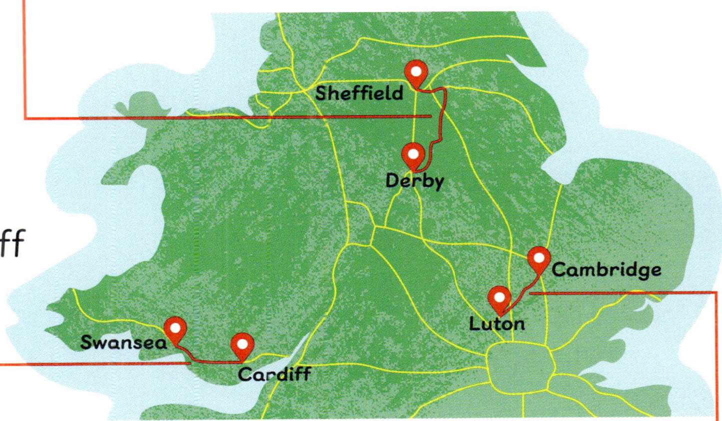

Luton to Cambridge
63 km

Is Ravi correct?

Master

We can use miles or kilometres to show distance.

To compare the distances, they all need to use the same unit of measurement.

1 Write the distance from Sheffield to Derby in kilometres. Round your answer to the nearest kilometres.

$1 \text{ mi} \approx 1.6 \text{ km}$

$49 \text{ mi} \approx 49 \times 1.6 \text{ km}$

$= 78.4 \text{ km}$

$\approx 78 \text{ km}$

$1 \text{ mi} \approx 1.6 \text{ km}$
Multiply by 1.6 to convert from miles to kilometres.

Sheffield to Derby is the longest distance.
Ravi is incorrect.

2 The distance from Luton to Reading is 57 mi.
Which is further from Luton: Reading or Cambridge?

Cambridge is 63 km away from Luton.

63 km ≈ ____ mi

1 km ≈ 0.62 mi
63 km ≈ 63 × 0.62 mi
= 39.06 mi
≈ 39 mi

57 mi > 39 mi
Reading is further from Luton than Cambridge.

Guided Practice

1 The table shows the driving distance between several locations.
Fill in the missing distances in miles or kilometres.
Round your answer to the nearest whole number.

Trip	Distance (mi)	Distance (km)
Gloucester to Hereford	28 mi	
Nottingham to Leicester		44 km
Birmingham to Derby	43 mi	
Norwich to Colchester		96 km
Hull to Wakefield	56 mi	

2 The longest driving distance in miles is ____ mi from ____ to ____.

3 The shortest driving distance in kilometres is ____ km from ____ to ____.

Complete Worksheet **4**: pages **141–142**

Converting Units of Mass

Lesson 5

Explore

Charles wants to bake some chocolate biscuits using this recipe.

1200 g	sugar	1680 g	flour
480 g	butter	240 g	cocoa powder

He has the following ingredients in his pantry.

1.5 kg of sugar
1 kg 80 g of flour
2.05 kg of butter
0.2 kg cocoa powder

Does Charles have enough of each ingredient?

Master

1 Write 1.5 kg in grams.

1 kg = 1000 g
1.5 kg = 1.5 × 1000 g
= 1500 g

1.5 g < 1200 g
Charles has enough sugar.

1 kg = 1000 g
Multiply by 1000 to convert from kilograms to grams.

2 Write 1 kg 80 g in grams.

$$\begin{aligned} 1\text{ kg} &= 1000\text{ g} \\ 1\text{ kg } 80\text{ g} &= 1000 + 80\text{ g} \\ &= 1080\text{ g} \end{aligned}$$

1 kg 80 g $<$ 1680 g
Charles does not have enough flour.

3 Write 2.05 kg in grams.

$$\begin{aligned} 1\text{ kg} &= 1000\text{ g} \\ 2.05\text{ kg} &= 2.05 \times 1000\text{ g} \\ &= 2050\text{ g} \end{aligned}$$

2.05 kg $>$ 480 g
Charles has enough butter.

4 Write 0.2 kg in grams.

$$\begin{aligned} 0.2\text{ kg} &= 0.2 \times 1000\text{ g} \\ &= 200\text{ g} \end{aligned}$$

0.2 kg $<$ 240 g
Charles does not have enough cocoa powder.

5 Charles' mum borrows 1 kg 200 g of flour from their neighbour. Write this amount in kilograms.

1 kg 200 g = ☐ kg

$$\begin{aligned} 1000\text{ g} &= 1\text{ kg} \\ 200\text{ g} &= 200 \div 1000\text{ kg} \\ &= 0.2\text{ kg} \\ 1\text{ kg } 200\text{ g} &= 1\text{ kg} + 0.2\text{ kg} \\ &= 1.2\text{ kg} \end{aligned}$$

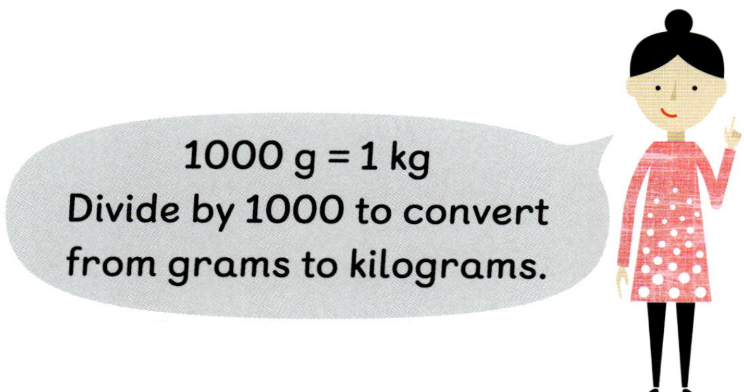

She borrows 1.2 kg of flour.

Guided Practice

1 Write in grams.

(a) 20 kg = ______ g

(b) 4.64 kg = ______ g

2 Write in kilograms.

(a) 35 g = ______ kg

(b) 500 g = ______ kg

3 Fill in the blanks.

(a) 1 kg 600 g = ______ g

(b) 4 kg ______ g = 4070 g

(c) ______ kg ______ g = 5028 g

4 Fill in the blanks.

(a) 9 kg ______ g = 9.62 kg

(b) ______ kg ______ g = 10.05 kg

(c) 15 kg 5 g = ______ kg

Complete Worksheet **5**: pages **143–144**

Converting Units of Volume

Lesson 6

Explore

Find the total volume of water Lulu, Elliott and Amira have altogether.

Master

1 Convert 1.1 l to millilitres and find the total.

1 l = 1000 ml
1.1 l = 1.1 × 1000 ml
= 1100 ml
300 ml + 550 ml + 1100 ml = 1950 ml

They have 1950 ml of water altogether.

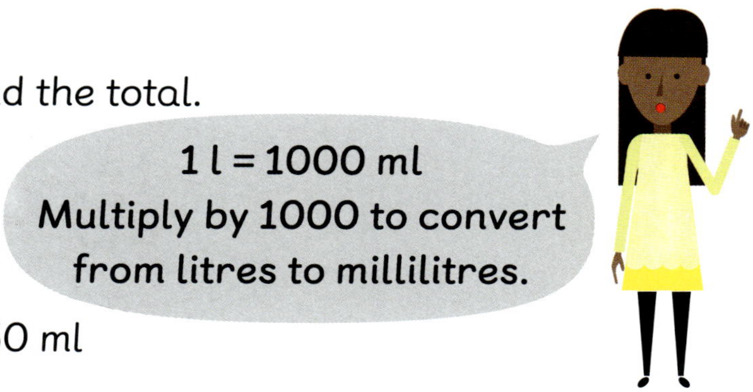

2 Convert 300 ml and 550 ml to litres and find the total.

1000 ml = 1 l
300 ml = 300 ÷ 1000 l
= 0.3 l

550 ml = 550 ÷ 1000 l
= 0.55 l

0.3 l + 0.55 l + 1.1 l = 1.95 l
They have 1.95 l of water altogether.

3 Express the total volume of water in litres and millilitres.

1.95 l = ___ l ___ ml
0.95 l = 0.95 × 1000 ml
= 950 ml
1.95 l = 1 l 950 ml

They have 1 l 950 ml of water altogether.

Guided Practice

1 Write in millilitres.

(a) 15 l = ___ ml

(b) 8.34 l = ___ ml

2 Write in litres.

(a) 303 ml = ___ l

(b) 10 750 ml = ___ l

3 Fill in the blanks.

(a) 5 l 200 ml = ___ ml

(b) 6 l ___ ml = 6008 ml

(c) ___ l ___ ml = 20.02 l

(d) 9 l 17 ml = ___ l

Complete Worksheet **6**: pages **145–146**

Lesson
7

Converting Units of Time

Explore

Ravi is taking the 1 p.m. train to York.
The trip will take 2.3 hours.
At what time will he arrive in York?

Master

1 Express 2.3 h in hours and minutes.

$1 \text{ h} = 60 \text{ min}$
$2.3 \text{ h} = 2 \text{ h} + 0.3 \text{ h}$
$0.3 \text{ h} = 0.3 \times 60 \text{ min}$
$= 18 \text{ min}$

$2.3 \text{ h} = 2 \text{ h } 18 \text{ min}$

2 h 18 min after 1 p.m. is 3:18 p.m.
Ravi will arrive in York at 3:18 p.m.

2 Express 2.3 h as a mixed number.

$0.3 = \frac{3}{10}$

$2.3 \text{ h} = 2 \text{ h} + \frac{3}{10} \text{ h}$

$= 2\frac{3}{10} \text{ h}$

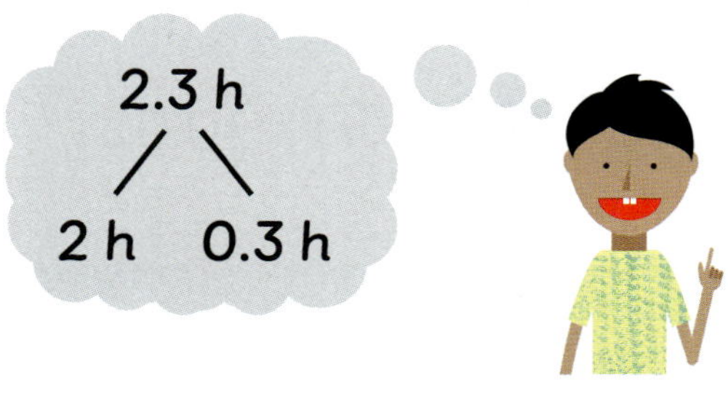

3 Ravi takes another train from York to Castleford. This trip takes 1 h 12 min. How long is this journey in h?

1 h 12 min = ___ h

60 min = 1 h

12 min = 12 ÷ 60 h

= 0.2 h

1 h 12 min = 1 h + 0.2 h = 1.2 h

The journey is 1.2 h long.

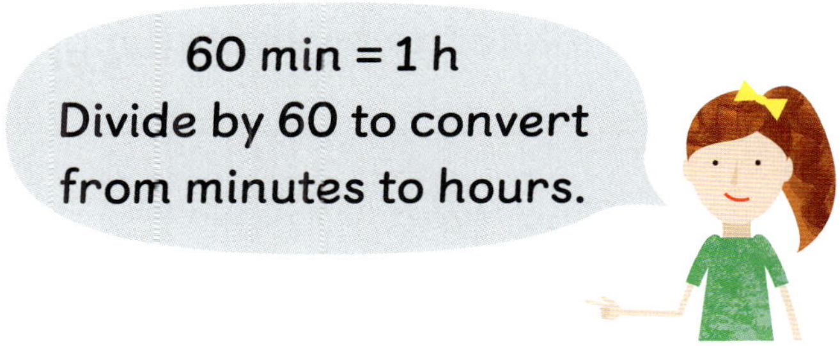

1 h 12 min = ___ h

12 min = $\frac{12}{60}$ h

= $\frac{1}{5}$ h

1 h 12 min = $1\frac{1}{5}$ h

The journey is $1\frac{1}{5}$ h long.

4 Ravi takes 4.8 minutes to walk to the bus station. How many minutes and seconds does he take?

1 min = 60 s

4.8 min = 4 min + 0.8 min

0.8 min = 0.8 × 60 s

= 48 s

4.8 min = 4 min 48 s

He takes 4 min 48 s to walk to the bus station.

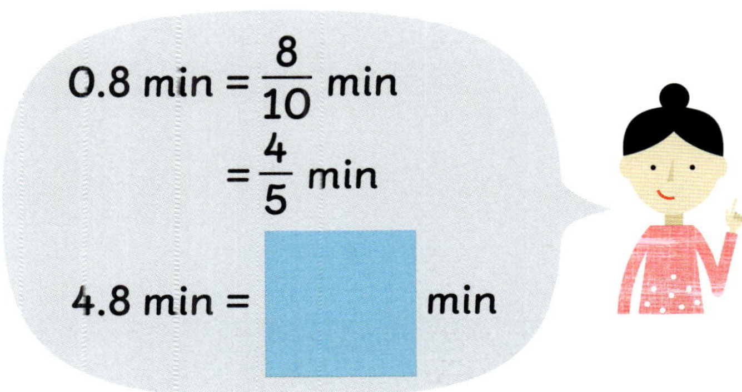

Guided Practice

1 Complete the table showing train journey times in hours and minutes.

Journey	Time (h)	Time (min)
London to Ely	1 h 10 min	☐ min
Harrogate to York	$\frac{☐}{☐}$ h	45 min
Oxford to Solihull	$1\frac{5}{6}$ h	☐ min
Tunbridge Wells to Hastings	$\frac{2}{3}$ h	☐ min
Hull to Whitby	☐ h ☐ min	200 min

2 The men's world record for the 800 m running event is $1\frac{2}{3}$ min while the women's record is 1 min 53 s.
How much longer is the women's record?
Give your answer in seconds.

Complete Worksheet **7**: pages **147–148**

Mind Challenge

Elliott's mum is driving from London to Birmingham, covering a distance of 128 mi. Her car uses 2 gallons of petrol every 100 km.
She has 3.5 gallons of petrol in the car. Will this be enough for her journey?

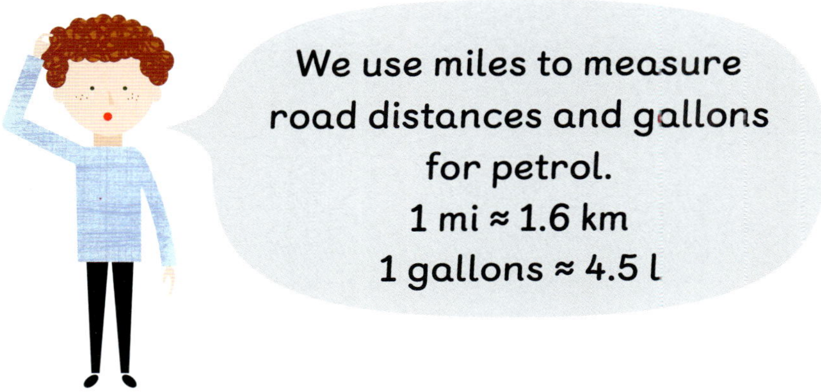

My Maths Journal

In the United Kingdom, we use both metric and imperial measurements. Find and record examples of both ways of measuring. Which system do you find the easiest to use and why?
Convert each measurement so that you have both the imperial and metric figures.

Review

I can...

- [ ] convert measurements of length.
- [ ] convert measurements of mass.
- [ ] convert measurements of volume.
- [ ] convert measurements of time.

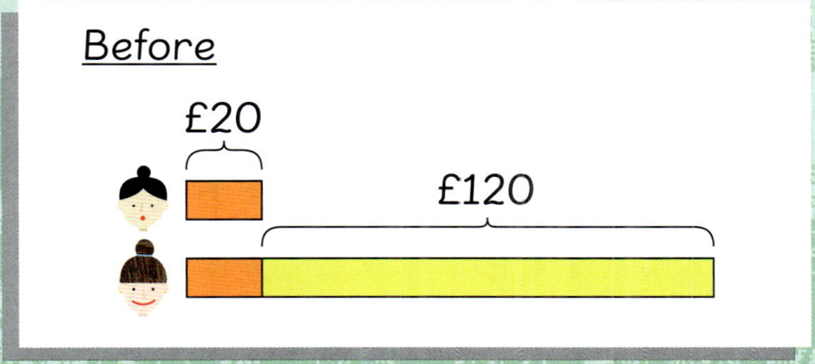

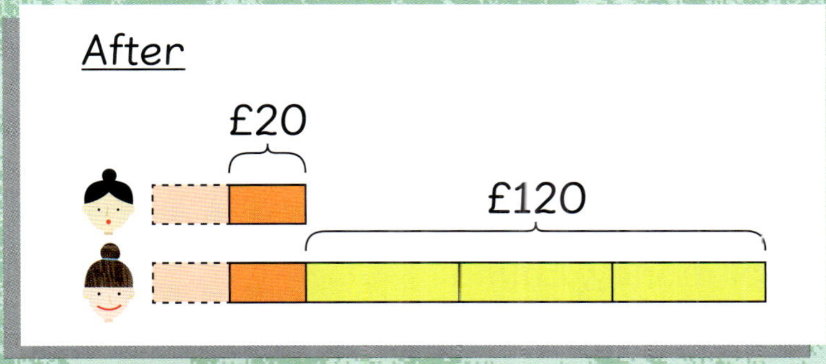

- ☐ Understand
- ☐ Plan
- ☐ Action the plan
- ☐ Check

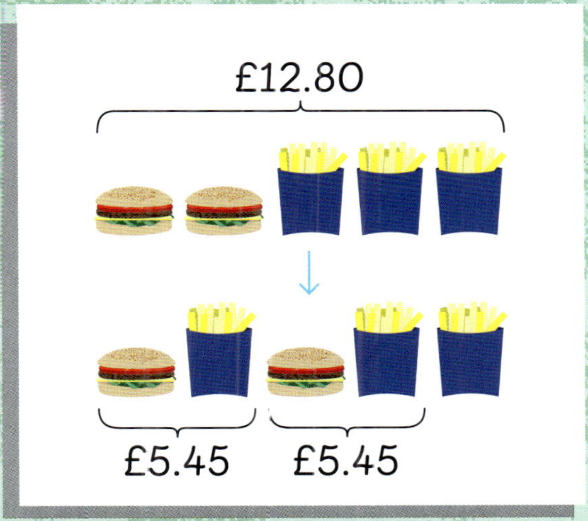

What are some of the different methods you would use when solving word problems?

Chapter 6
Word Problems

Solving Word Problems

Lesson 1

Explore

Ruby had £20. Emma had £120 more than Ruby. After Ruby's mum gave them both the same amount of money, Emma had four times as much money as Ruby. How much money did Ruby's mum give to each girl?

Master

1 **Understand**

Before

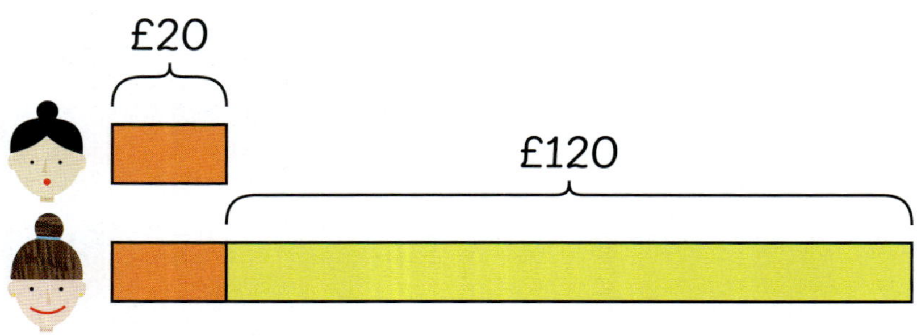

- [ ] Understand
- [ ] Plan
- [ ] Action the plan
- [ ] Check

Ruby's mum gives both girls the same amount.

After

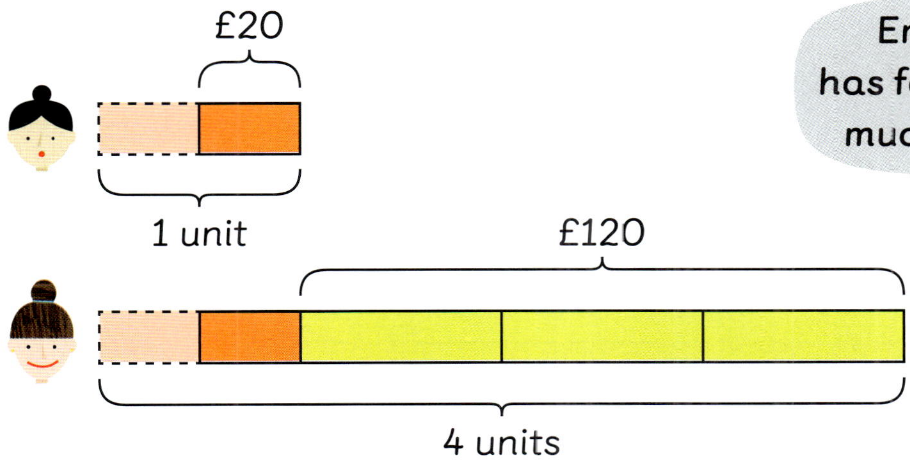

Plan

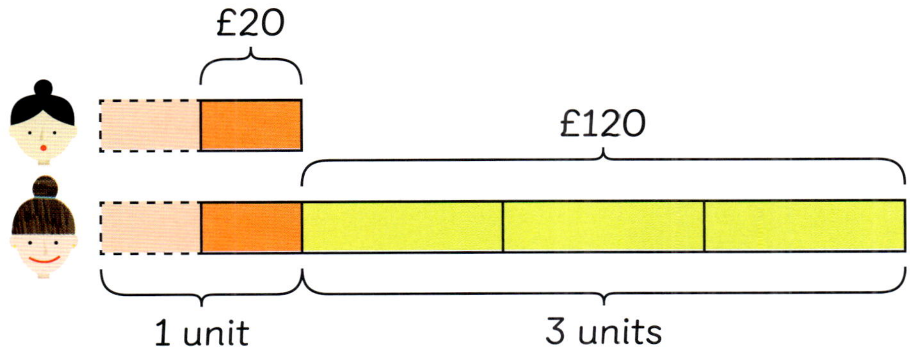

Action the plan

3 units = £120

1 unit = £120 ÷ 3

= £40

 = £40 − £20

= £20

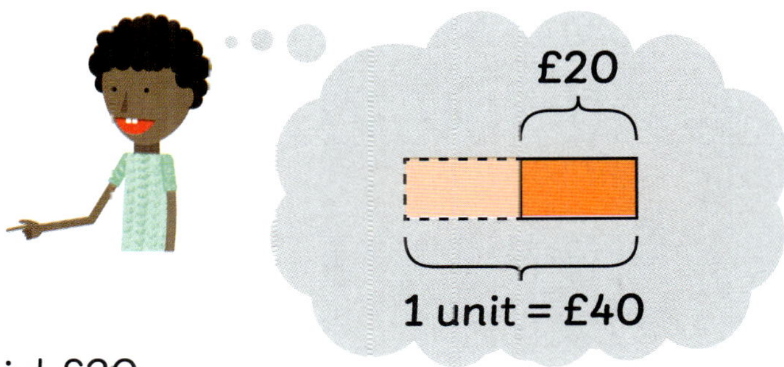

Ruby's mum gave each girl £20.

Check

Before	Ruby had £20. Emma had £140.	After	Ruby has £40. Emma has £160.

£160 = 4 × £40

2 Emma, Elliott and Ravi record the distance they cycle on their bikes. Emma cycles four times the distance that Ravi cycles. Elliott cycles 5 km further than Ravi cycles. Altogether, they cycle a total distance of 227 km.
What is the distance Emma cycles?

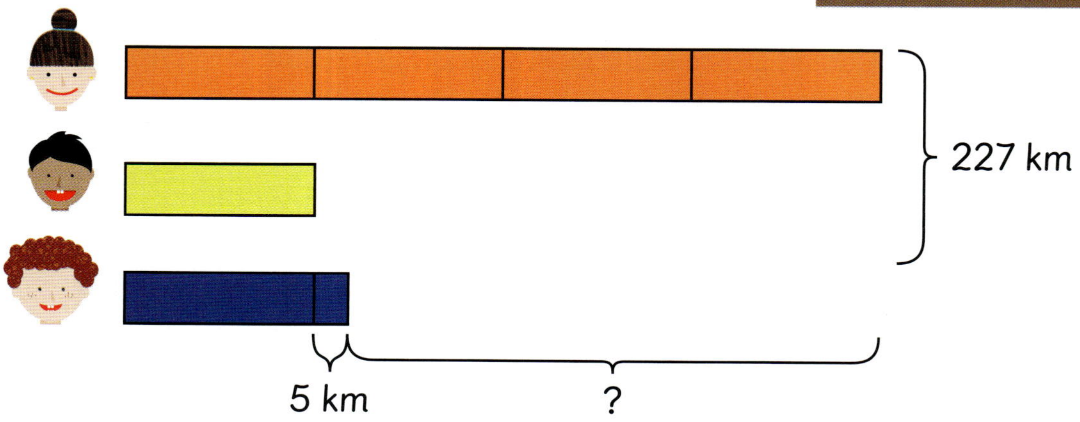

227 − 5 = 222

6 units = 222 km
1 unit = 222 km ÷ 6
= 37 km

Distance Emma cycles = 37 km × 4
= 148 km

Emma cycles 148 km.

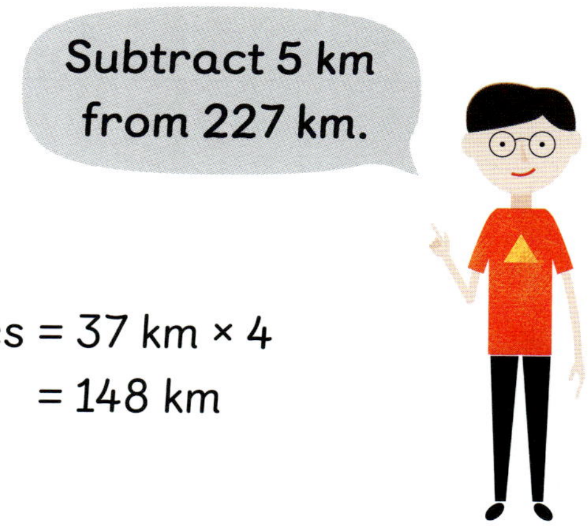

- ☐ Understand
- ☐ Plan
- ☐ Action the plan
- ☐ Check

Guided Practice

1 Amira, Lulu and Elliott share £108. Amira has three times as much money as Elliott and Lulu has twice as much as Elliott. How much money do they each have?

2 A watermelon is twice as heavy as a cantaloupe.
A bag of potatoes is 1.5 kg heavier than the cantaloupe.
The total mass of the watermelon, the cantaloupe and the bag of potatoes is 9.5 kg.

(a) Find the mass of the watermelon.

(b) Find the mass of the bag of potatoes.

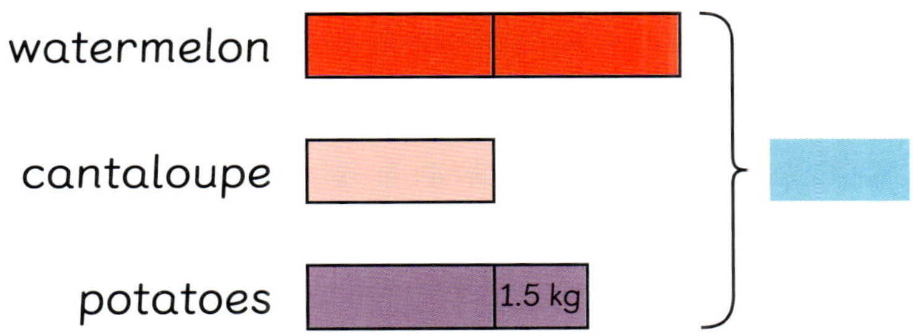

3 At first, Oak had £87 and Charles had £31. After they each spent an equal amount of money, Oak had three times as much money as Charles. How much did each of them spend?

Complete Worksheet **1**: pages **153–154**

Solving Word Problems

Lesson 2

Explore

Holly bought 6 sandwiches. Sam bought 4 sandwiches. Sam also bought 3 drinks at £1.30 each. Altogether, Sam spent £2.50 less than Holly. How much did each sandwich cost?

Master

1 How much did each sandwich cost?

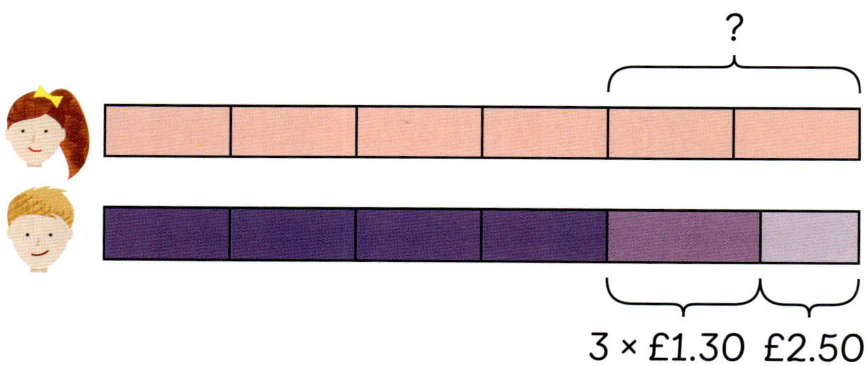

- ☐ Understand
- ☐ Plan
- ☐ Action the plan
- ☐ Check

3 × £1.30 = £3.90
£3.90 + £2.50 = £6.40

$$\begin{array}{r} \pounds\ {}^{1}3.90 \\ +\ \pounds\ 2.50 \\ \hline \pounds\ 6.40 \\ \hline \end{array}$$

2 units = £6.40
1 unit = £3.20

Each sandwich cost £3.20.

2 How much did Holly and Sam pay altogether?

Method 1

 £3.20 × 6 = £19.20

 £3.20 × 4 + £1.30 × 3 = £ ___

£19.20 + £ ___ = £ ___

They paid £ ___ altogether.

Method 2

£3.20 × 10 + £1.30 × 3
= £30 + £2 + £3.90
= £32 + £3.90
= £35.90

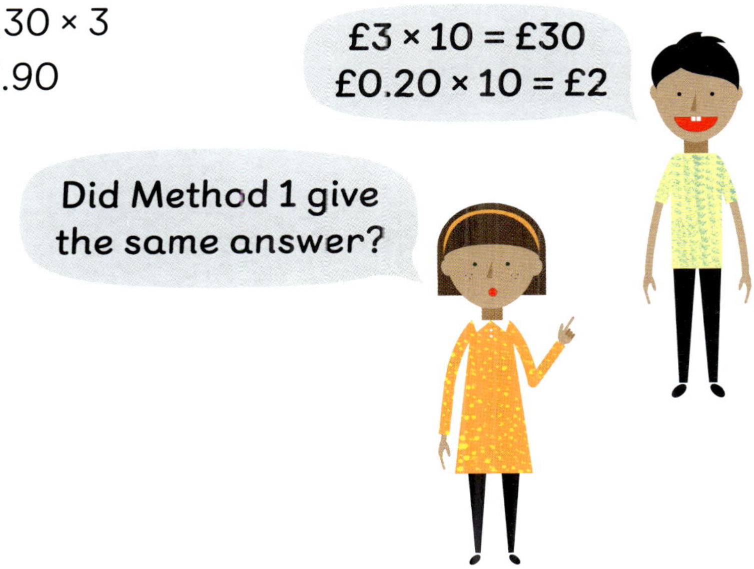

3 How much change will they get using four £10 notes?

4 × £10 = £40

	£	~~4~~ 2	~~0~~ 10 9	.	10 0
−	£	3	5	.	9 0
	£		___	.	___ ___

£40 − £35.90 = £ ___

They will get £ ___ in change.

Guided Practice

1 2  and 3  costs £12.80.

1  and 1  costs £5.45.

What would 4  and 2  cost?

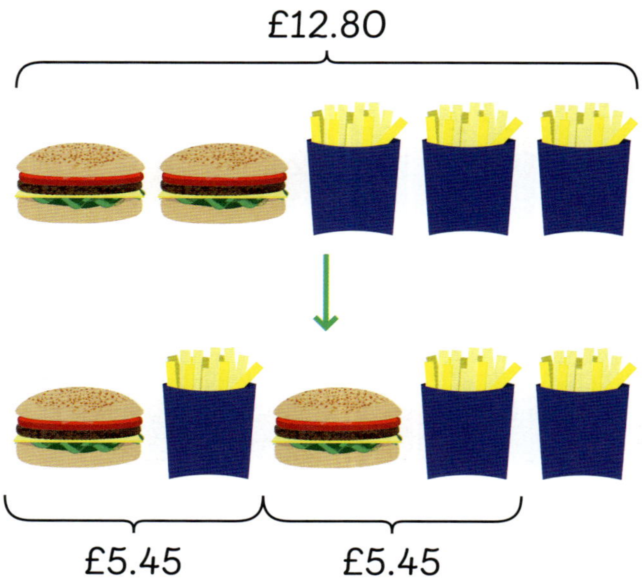

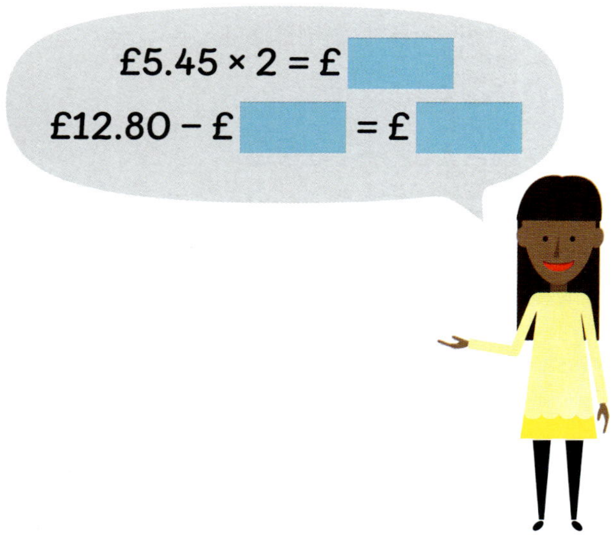

2 5 laptops weigh 5.4 kg more than 2 laptops and 2 tablets. Each tablet has a mass of 0.6 kg. What is the mass of each laptop?

3 The cost of 3 magnets and 4 bells is £47. The cost of 3 magnets and 6 bells is £63. What is the cost of 1 magnet?

Complete Worksheet **2**: pages **155–156**

Lesson 3

Solving Word Problems

Explore

The mass of 2 parcels is the same as the mass of 5 bags. The total mass of 6 parcels and 3 bags is 10.8 kg.
Find the total mass of 1 parcel and 1 bag.

Master

mass of 2 parcels = mass of 5 bags

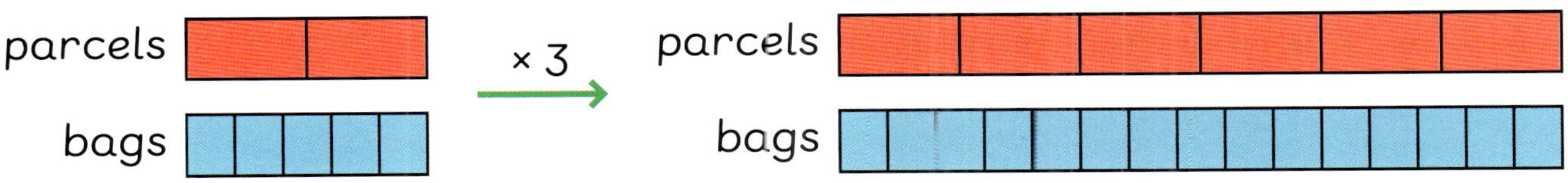

mass of 6 parcels = mass of 15 bags

- ☐ Understand
- ☐ Plan
- ☐ Action the plan
- ☐ Check

mass of 6 parcels + mass of 3 bags
= mass of 15 bags + mass of 3 bags
= mass of 18 bags

mass of 18 bags = 10.8 kg
mass of 1 bag = 10.8 ÷ 18 = 0.6 kg

108 tenths ÷ 18 = 6 tenths

mass of 2 parcels = mass of 5 bags
= 0.6 × 5
= 3 kg

mass of 1 parcel = 3 kg ÷ 2
= 1.5 kg

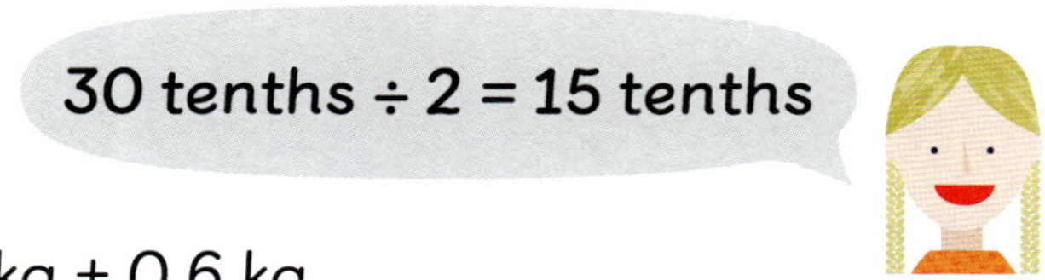

mass of 1 parcel and 1 bag = 1.5 kg + 0.6 kg
= 2.1 kg

The total mass of 1 parcel and 1 bag is 2.1 kg.

Guided Practice

1 3 pencil cases cost as much as 7 pens. Each pencil case costs £1.20 more than each pen. What is the cost of a pencil case?

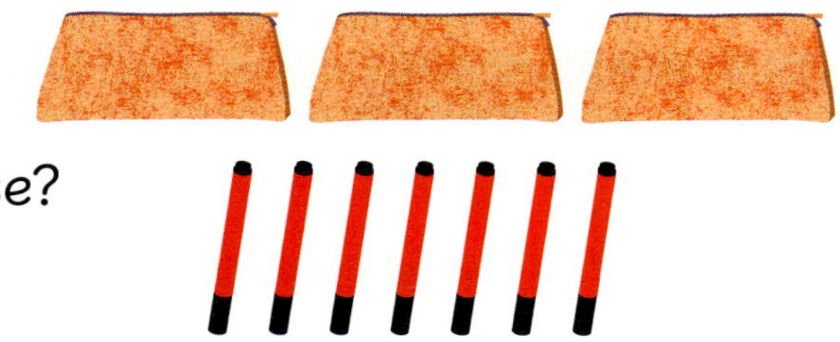

2 2 bars of chocolate cost £4 more than 3 slices of cake. Holly bought 8 bars of chocolate and 7 slices of cake for £54. What was the cost of a bar of chocolate?

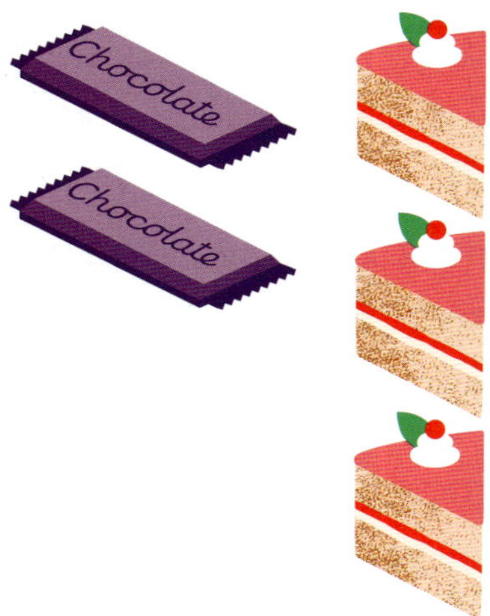

Complete Worksheet **3**: pages **157–158**

Solving Word Problems

Lesson 4

Explore

Ravi and Hannah went for a hike with their parents. The trail they hiked on was divided into three sections. It took them three times as long to hike Section 2 than Section 1. It took them 10 minutes less to hike Section 3 than Section 2. The entire hike took them 1 hour 35 minutes.

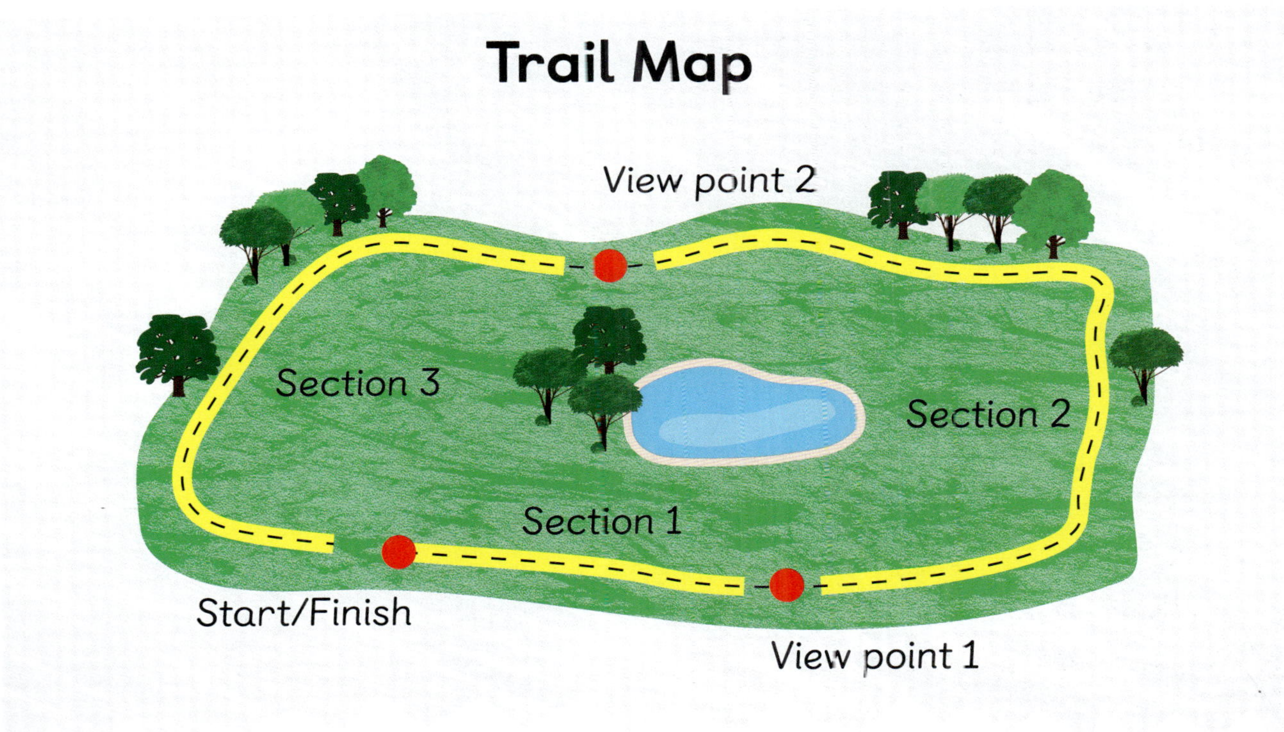

How long did it take them to complete Section 3?

Master

We can draw bar models.

1 h = 60 min
1 h 35 min = 95 min

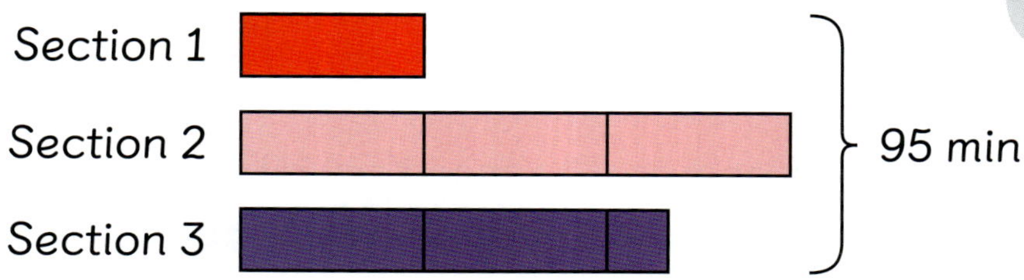

95 min + 10 min = 105 min

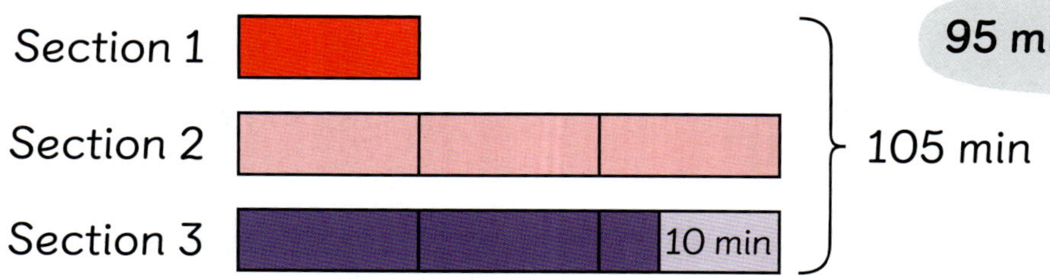

7 units = 105

1 unit = 105 ÷ 7 = 15

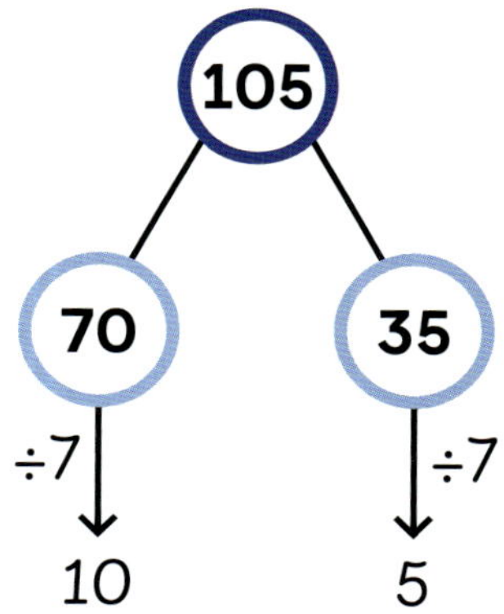

10 + 5 = 15

3 units = 3 × 15
= 45

45 − 10 = 35

It took them 35 minutes to complete Section 3.

- ☐ Understand
- ☐ Plan
- ☐ Action the plan
- ☐ Check

Guided Practice

1 It takes Lulu 20 minutes to get ready. She takes five times as long to take a shower than to brush her teeth. It then takes her 2 minutes to get dressed. How long does Lulu brush her teeth for?

2 A lorry driver completed three deliveries in one morning. The first delivery took twice as long as the second delivery. The third delivery took 5 minutes more than the first delivery. The three deliveries took 2 hours 35 minutes in total. How long did the third delivery take?

3 Write a word problem that can be solved using this bar model.

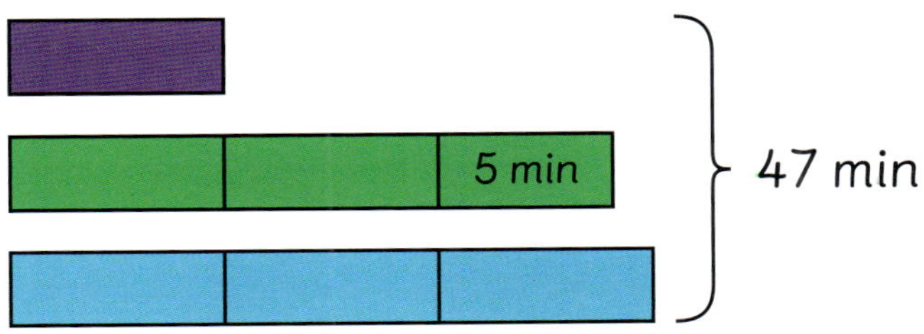

Complete Worksheet **4**: pages **159–160**

Solving Word Problems

Lesson 5

Explore

An animal shelter for cats and dogs currently houses 240 animals.

$\frac{1}{2}$ the number of dogs is equal to $\frac{3}{4}$ of the number of cats.

How many cats are there?

Master

1

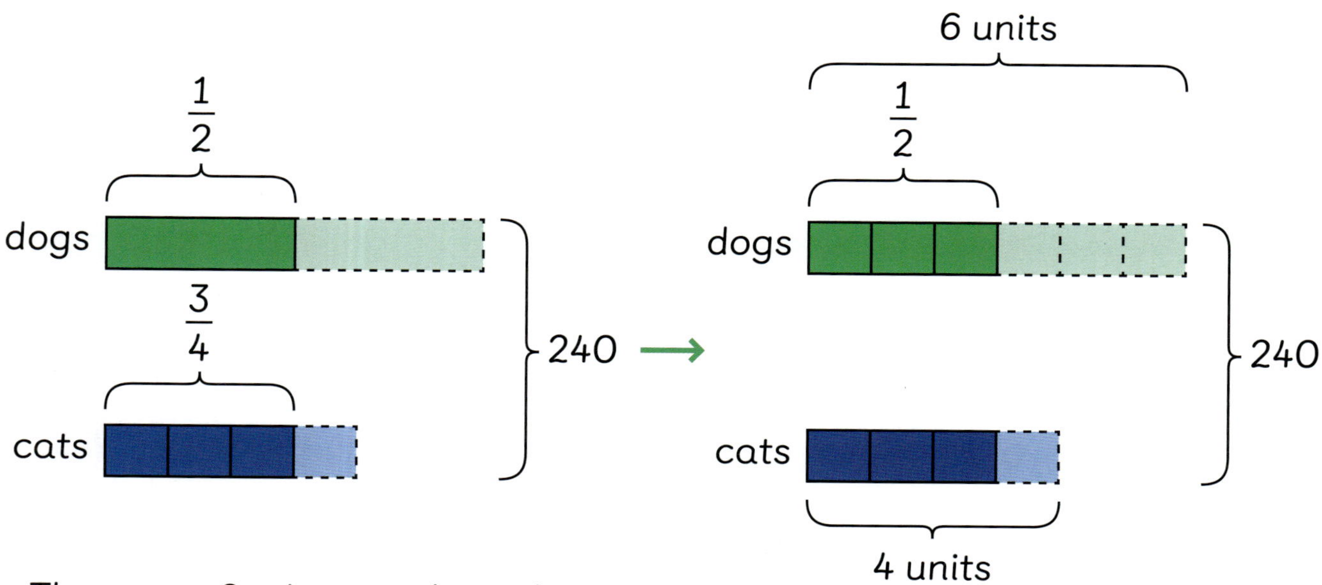

There are 6 + 4 units altogether.

10 units = 240

1 unit = 24

$6 \times 24 = 144$
There are 144 dogs.

$4 \times 24 = 96$
There are 96 cats.

2 Charles spends $\frac{2}{5}$ of his pocket money on comic books. Ruby spends $\frac{4}{7}$ of her pocket money on playing cards. They each spend the same amount of money. How much money did each of them start with if they have £18 remaining in total?

Spent

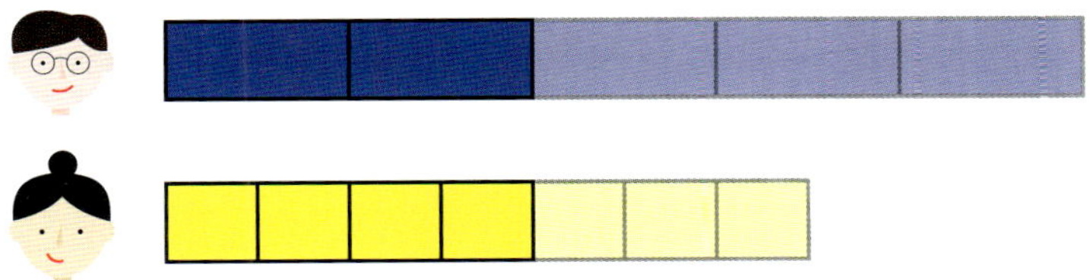

Remaining

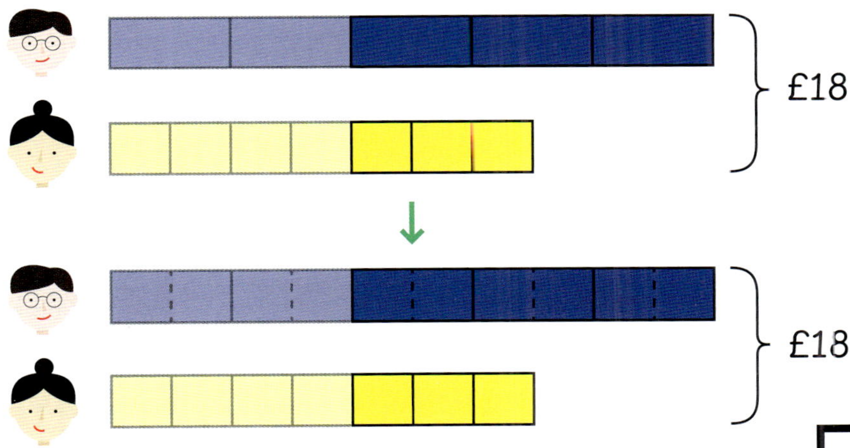

9 units = £18
1 unit = £18 ÷ 9
= £2

Charles → $10 \times £2 = £20$
Ruby → $7 \times £2 = £14$

Charles started with £20.
Ruby started with £14.

- ☐ Understand
- ☐ Plan
- ☐ Action the plan
- ☐ Check

Guided Practice

1 A bag contains 180 red and blue counters. $\frac{1}{3}$ of the red counters is equal to $\frac{1}{2}$ of the blue counters. How many of each type of counters are there?

2 Oak uses $\frac{3}{5}$ of a bag of white flour to bake a loaf of bread. She uses $\frac{6}{7}$ of a bag of brown flour to bake another loaf of bread. She uses an equal mass of flour for both loaves. After baking the loaves, Oak has 150 g of white and brown flour left altogether. What was the total mass of white flour in the bag before Oak baked the loaf of bread?

Complete Worksheet **5**: pages **161–162**

Solving Word Problems

Lesson 6

Explore

Hannah, Charles and Jacob each write a word problem for the calculation 32 ÷ 5.
Hannah's word problem gives the answer 2.
Charles' word problem gives the answer 6.
Jacob's word problem gives the answer 7.
Can they all be correct?

Master

1 32 ÷ 5

A farmer wants to share his 32 sheep equally between 5 pens. How many sheep will be left over?

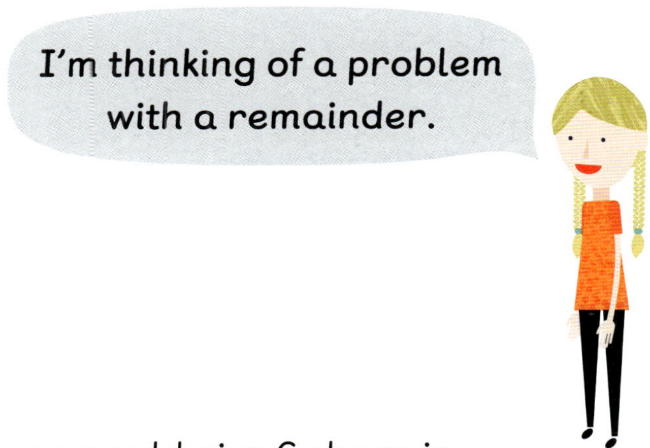

6	6	6	6	6	2

32 sheep shared equally between 5 pens would give 6 sheep in each pen with 2 sheep left over.

2 32 pencils are shared equally between 5 children. How many pencils do they each receive?

$32 \div 5 = 6$ remainder 2

6	6	6	6	6	2

Each child receives 6 pencils.

3 32 children will travel in cars for a school trip. Each car can hold up to 5 children. What is the least number of cars that will be needed for all the children to go on the trip?

5	5	5	5	5	5	2

7 cars will be needed for all the children to go on the trip.

Guided Practice

1 105 pupils are going to an overnight school camp. Each tent can only sleep 4 pupils. What is the smallest number of tents needed for all the pupils?

2 Amira uses 40 ml of syrup to make a milkshake. What is the greatest number of milkshakes she can make with 500 ml of syrup? How much syrup will be left over?

Complete Worksheet **6**: pages **163–164**

Mind Challenge

Ravi used $\frac{2}{5}$ of his tokens on an online game. After Holly used $\frac{2}{5}$ of her tokens, she had $\frac{1}{2}$ the number of tokens that Ravi had left.

If Holly used 8 tokens, how many tokens did Ravi start with?

My Maths Journal

Use the words in the boxes below to write a word problem, then solve it.

255 roses	a dozen roses	packs	boxes

Review

I can...

☐ solve word problems using multiple methods.